THE TRIUMPH OF THE SUN

Pan Stanford Series on Renewable Energy

Series Editor

Wolfgang Palz

Vol. 1
Power for the World: The Emergence of Electricity from the Sun
Wolfgang Palz, ed.
2010
978-981-4303-37-8 (Hardcover)
978-981-4303-38-5 (eBook)

Vol. 2
Wind Power for the World: The Rise of Modern Wind Energy
Preben Maegaard, Anna Krenz, and Wolfgang Palz, eds.
2013
978-981-4364-93-5 (Hardcover)
978-981-4364-94-2 (eBook)

Vol. 3
Wind Power for the World: International Reviews and Developments
Preben Maegaard, Anna Krenz, and Wolfgang Palz, eds.
2013
978-981-4411-89-9 (Hardcover)
978-981-4411-90-5 (eBook)

Vol. 4
Solar Power for the World: What You Wanted to Know about Photovoltaics
Wolfgang Palz, ed.
2013
978-981-4411-87-5 (Hardcover)
978-981-4411-88-2 (eBook)

Vol. 5
Sun above the Horizon: Meteoric Rise of the Solar Industry
Peter F. Varadi
2014
978-981-4463-80-5 (Hardcover)
978-981-4613-29-3 (Paperback)
978-981-4463-81-2 (eBook)

Vol. 6
Biomass Power for the World: Transformations to Effective Use
Wim van Swaaij, Sascha Kersten, and Wolfgang Palz, eds.
2015
978-981-4613-88-0 (Hardcover)
978-981-4669-24-5 (Paperback)
978-981-4613-89-7 (eBook)

Vol. 7
The U.S. Government & Renewable Energy: A Winding Road
Allan R. Hoffman
2016
978-981-4745-84-0 (Paperback)
978-981-4745-85-7 (eBook)

Vol. 8
Sun towards High Noon: Solar Power Transforming Our Energy Future
Peter F. Varadi
2017
978-981-4774-17-8 (Paperback)
978-1-315-19657-2 (eBook)

Vol. 9
The Sun Is Rising in Africa and the Middle East: On the Road to a Solar Energy Future
Peter F. Varadi, Frank Wouters, and Allan R. Hoffman
2018
978-981-4774-89-5 (Paperback)
978-1-351-00732-0 (eBook)

Vol. 10
The Triumph of the Sun: The Energy of the New Century
Wolfgang Palz
2018
978-981-4800-06-8 (Hardcover),
978-0-429-48864-1 (eBook)

Pan Stanford Series on Renewable Energy – Volume 10

THE TRIUMPH OF THE SUN

The Energy of the New Century

Wolfgang Palz

Published by

Pan Stanford Publishing Pte. Ltd.
Penthouse Level, Suntec Tower 3
8 Temasek Boulevard
Singapore 038988

Email: editorial@panstanford.com
Web: www.panstanford.com

British Library Cataloguing-in-Publication Data
A catalogue record for this book is available from the British Library.

The Triumph of the Sun: The Energy of the New Century

ISBN 978-981-4800-06-8 (Hardcover)
ISBN 978-0-429-48864-1 (eBook)

A tribute to

my good friend, the late Hermann Scheer, MP

Alternative Nobel Prize Winner (1999)

Hero of the Green Century (US Time Magazine, *2002*)

A Tribute to the Glory of the Sun

Strauss, Richard. "Sonnenaufgang" (Sunrise), *Also sprach Zarathustra*. Perf. Herbert von Karajan and the Berliner Philharmoniker, 1984, Berlin.

https://www.youtube.com/watch?v=6Hi5xbguTJk.

Also sprach Zarathustra is a symphonic poem by Richard Strauss, composed in 1896 and inspired by Friedrich Nietzsche's philosophical novel of the same name. The first part of the poem is called *Einleitung, oder Sonnenaufgang* (Introduction, or Sunrise).

Stravinsky, Igor. *The Firebird*, Perf. Pierre Boulez and Orchestre de Paris, 2009, Paris.

https://www.youtube.com/watch?v=pTbwQ6G-bP0.

The Firebird is a ballet and orchestral concert work by the Russian composer Igor Stravinsky. It was written for the 1910 Paris season of Sergei Diaghilev's company Ballets Russes; the original choreography was by Michel Fokine, with a scenario by Alexandre Benois based on the Russian fairy tales of the Firebird and the blessing and curse it possesses for its owner.

van Beethoven, Ludwig. *Sonata No. 21*. Perf. Emil Gilels, piano, 1971, Ossiach.

https://www.youtube.com/watch?v=5U0LWqMPU20.

Beethoven's *Piano Sonata No. 21 in C Major, Op. 53*, known as the *Waldstein*, is one of the three most notable sonatas of his middle period. It is also known as *L'Aurora* (The Dawn) in Italian, for the sonority of the opening chords of the third movement, thought to conjure an image of sunrise—daybreak.

Paganini, Niccolò. *La Campanella*. Perf. Clara Jumi Kang, violin, 2015, Saint Petersburg.

https://www.youtube.com/watch?v=42O0EZkeQ_c.

The *Violin Concerto No. 2 in B Minor, Op. 7*, was composed by Niccolò Paganini in Italy in 1826. In his Second Concerto, Paganini holds back on the demonstration of virtuosity in favor of greater individuality in the melodic style. The third movement of Paganini's Second Concerto owes its nickname "La Campanella" or "La Clochette" to the little bell which Paganini prescribes to presage each recurrence of the rondo theme.

Mozart, Wolfgang Amadeus. *Piano Concerto No. 23*. Perf. Armen Manassian, piano, 2013, Moscow.

https://www.youtube.com/watch?v=qpT7XDWhiA4.

This concerto in A major is a composition for piano and orchestra written by Mozart. It was finished, according to Mozart's own catalogue, on 2 March 1786, two months prior to the premiere of his opera, Le nozze di Figaro. It was one of three subscription concerts given that spring and was probably played by Mozart himself at one of these.

Brahms, Johannes. *Piano Quintet Op. 34*. Perf. Quatuor Simon and Ionel Streba, 2014, Paris.

https://www.youtube.com/watch?v=RPmKKqX5xV0.

The quintet in F minor was completed by Brahms during the summer of 1864 and published in 1865. It was dedicated to Her Royal Highness Princess Anna of Hesse. The work, "often called the crown of his chamber music," began life as a string quintet. Brahms transcribed the quintet into a sonata for two pianos (in which form Brahms and Carl Tausig performed it) before giving it its final form.

Note: All annotations are from Wikipedia. Readers can listen to these works by following the given links.

Music selection: Courtesy of ARTCONCEPT INTERNATIONAL ASSOCIATION

Contents

PART 3: UNDERSTANDING NATURE, CREATING KNOW-HOW

About the Author

Dr. Wolfgang Palz is a German physicist. For over 50 years, he has been one of the global leaders of the development of solar energy and all the renewables. After obtaining a PhD from the University of Karlsruhe, Germany, he became an official at the French National Space Administration (CNES) with responsibility for the development of solar PV. In 1977, he became head of division at the EU Commission in charge of the development of solar energy and renewables for Europe and the world. He kept that position for 20 years before moving to the EU Commission's Directorate of Aid. From 2000 to 2002, he was a member of an "Enquête Commission" of the German Parliament on the German energy perspective on the horizon 2050. Later, he worked under consultancy with the EU Commission on PV programmes for the poor in Latin America. He was also involved in various French national programmes, including one on investments for the future, which was just recently completed. Dr. Palz has received numerous awards, including the International Solar Energy Society (ISES) Global Leadership Award in Advancing Solar Energy Policy (2011). He is a bearer of an Order of Merit of the German Republic.

Preface

With the turn of the century in 2000, the world started to turn its back on the wrong-headed developments of the past with global pollution and the misery it entails, a climate getting out of control, the threat of a nuclear war, all of which was a result of the unsustainable use of fossil and atomic resources.

Not everybody may have realised it, with the 21st century, we have resolutely engaged ourselves again on the route towards a life in harmony with nature, with the Sun. This book is not about ecological dreams and wishful thinking for a better world. It is simply a report about what happened, in facts and figures.

Going definitely now with the Sun and its benefits, everybody is a winner, not only the climate. Thanks to innovation and mass production, the power derived from the Sun now beats the conventional world with its own strength: socio-economy. In our new world, solar energy has become cheaper than the conventional ones. We got a booming economy that is sustainable, with millions of new jobs for everyone.

The book starts from fundamentals and discusses the key role of the Sun for nature and our lives. It reports what happened when the foundation for a cleaner world was laid towards the beginning of the new century, detailing the efforts of the people who brought about the change.

The book is dedicated to a key figure who spearheaded this change to a better world, a solar world: the late Hermann Scheer.

The author, Wolfgang Palz, is an independent expert on energy matters and the economy. The book provides a summary of his global views on a solar revolution to which he contributed, his satisfaction that eventually the pioneers' aspirations were crowned with success.

Acknowledgements

The author expresses his sincere thanks to Arvind Kanswal of Pan Stanford Publishing for his tremendous contribution in the preparation of this book. The author is also grateful to his friend Peter Varadi for his continuous encouragement.

Prologue: A Vision of the Future from the 1970s

Nuclear power and coal had their heyday in the second half of the last century. In particular, since the oil-price shock of 1973, industrialised countries were anxious to preserve energy independence. Hundreds of nuclear power stations were set up in a record time.

Yet solar electricity and wind power had their adepts looking back to a long tradition in Europe. Interest had arisen in particular in the administration of the United States, too. President Carter did his best to support the new solar technologies, but in vain. While hundreds of Gigawatts of new atomic power was installed around the globe, solar PV was kept down at best to a few megawatts. "Too expensive," they said.

In 1977–1978, I published with UNESCO in Paris the book *Solar Electricity: An Economic Approach to Solar Energy*. It intended to summarise the understanding and mood of the solar experts in the field in Europe and the United States. The US administration had done a lot of investigations. A "Project Independence Report" had been looking in all detail into the prospects of the renewables. But it was kept unpublished for the general public. Hence, the book I published with UNESCO in English and a few other languages was for many a first encounter with solar energy. Following are a few excerpts of that farsighted book.

"There is only one way to diminish the various types of pollution brought about by man's large-scale consumption of energy, namely: direct use of the energy that dominates Earth's climate. Useful energy can be produced from Solar radiation in such a way that neither thermal nor chemical pollution whatsoever is caused".

The author's book on solar energy published in 1977 with UNESCO in Paris.

"All the known ways in which the Sun's radiation can be converted into useful power are discussed. Attention is focused on the direct conversion of light into electricity by means of Solar cells".

"The energy available in the form of Solar energy is evenly distributed. Thus, every country owns more potential energy it would ever need, renewed every year by the Sun. Solar energy is a homemade reserve".

"The development of Solar energy applications does not mean the beginning of a new economic world. On the contrary, the new energy systems must first win their place in the overall energy market, they must be made competitive with oil, coal, or nuclear energy, whether for reasons of depletion of conventional resources, thermal or chemical pollution of the natural environment, greater independence from foreign suppliers or simply lower cost".

"The technical and economic problems associated with the large-scale use of Solar energy are explored". "Assessment of Solar energy's large-scale potential for the future: Evidence is given that the "present" high cost of solar cells is by no means inevitable and that a large-scale reduction of manufacturing costs down to the level required for cost-effective central power plants can be expected in the next 10 to 15 years". "Economy of scale". "Progress in industrialisation".

"In 1975 PV terrestrial market was only 100 kW against a yearly installation rate of conventional power of hundreds of MW".

"PV large-scale production volume of 10 GW leads to a cost of $0.20 to $0.50". "The cumulative production volumes associated with a reasonable learning curve can in fact be achieved".

"If central PV power plants are integrated in an extensive power grid no special problem will occur since the situation is the same as for conventional power plants". "Only as an independent power generator it is preferable to add an electrical storage device".

"PV power generators employing very low-cost Solar cells will be cost effective at almost any power level, even at some Watts or kW. Thus it is possible to envisage individual generators for homes, community plants for villages, shopping centres, industrial production plants, agricultural processing and farms—as well as central power plants".

"Solar generators installed close to the consumers may prove to be attractive because they avoid excessive transmission costs, and when mounted on roof tops or other available structures eliminate the need for land purchasing, site preparation, and supports".

"An array of 45 m^2 would fit on the roof of most family houses in the United States. If a lead acid battery is used for storage it would have a capacity of about 200 kWh, its volume would be 4 m^3. Such a system would give complete autonomy to the house".

Prologue: From the Triumph of the Iron to the Triumph of the Sun

The Triumph of the Iron

This was the motto of the big "Exposition Universelle de Paris" in 1889—with the brand new Eiffel Tower standing proudly in the middle. It actually was the year zero of the world's development that we have seen since then.

The first automobiles came to the roads in Germany at that time, and just a few years earlier, Thomas Edison had started operation of his first electric power plants in England and the United States. By the way, he also attended that World Exhibition in Paris, the City of light. Can you imagine today a world without electricity and without automobiles? That was the time before 1889, on the doorsteps of the 20th century.

The explosive growth of electricity supply that followed entailed an equally explosive growth of the consumption of dirty coal to feed the hundreds of new power plants. Things degraded further when in the 1970s atomic power plants got the favours of the politicians: Nuclear was in those days "unlimited amounts of energy for free". Four hundred of them have been built and installed until the turn of the century in just 30 years. When one looks back, it appears like a nightmare.

Forbes Magazine wrote in 1985, "The failure of the US nuclear power ranks as the largest managerial disaster in business history, a disaster on a monumental scale—only the blind or the biased can now think that the money was spent well".

In the last century, a few other things went wrong as well. Two World Wars with millions of innocent deaths, two dictatorships bringing misery and death to more millions of people, a world economic crisis with disastrous consequences. The world's population grew to an almost unsustainable size, and pollution of the natural environment, the air, the ground, and the seas affected the world on a scale never seen before, with climate change only one of the consequences. Mountains of millions of tonnes of plastic waste, the air in many cities around the world hardly breathable. Has the 20th century started an Anthropocene?

The Eiffel Tower in Paris (picture by the author).

The new century eventually brought a radical change. The discredit of solar energy and the renewables came to an end. What finished the nuclear option off was the explosion of two plants in Ukraine and Japan. Was it criminal in the first place to start running all those plants while nobody had the slightest idea where to dispose of all the dangerous nuclear waste produced, the explosions were too much and became the final straw for nuclear. As far as coal was concerned, it had its markets growing further into this century. However, eventually the pollution it entails and the risk of climate change for which it is most responsible put its development also to a standstill more recently.

Since the year 2000, the global production of electricity from nuclear plants has been turning down. In Europe, no new nuclear plant was put into operation in this century. The same holds for the United States, where, except the one that came online in 2016, several old plants were, by contrast, disconnected from the net. India and China put a few new ones in operation, but the expected revival of nuclear desperately expected by its supporters just did not take place. The world's nuclear industry—Areva, Westinghouse, Toshiba—is virtually bankrupt.

The world's consumption of coal has well made inroads into our new century; it has doubled since 1990. However, since 2013 it is no more increasing but stabilising. Since that year, the consumption of coal in US power plants stopped its growth. In 2016, it dropped to the level last seen in the 1970s. In the United Kingdom, coal output has fallen 82% between 2013 and 2017. China is the world's leader in coal consumption and it operates three times more coal capacity than the United States: However, China burned in 2016 the least amount of coal in 3 years. Bloomberg, the financial information provider, noted that it is the end of the era of coal: "Coal production is in freefall".

Together with the world's stabilisation of coal burning, CO_2 emissions have been stabilising for the past few years as well. Climatologists will like it.

In 2018, with only 18 years of age, our new century has just hardly become adult. And already it has swung the door open to an industrial revolution, the solar age. Since the turn of the century, we have indeed seen **the Triumph of the Sun**.

The Triumph of the Sun

It all started in Germany, Europe's largest economy. Why and how? We are going to see later. It would be worth a book on its own. In the year 2000, solar PV together with wind power and a few other renewables started there a breathtaking development. By 2017, electricity generation from the renewables has increased 10-fold to 38% of the total German consumption. Wind and PV now produce more electricity than coal and nuclear there.

In the European Union, more renewable power capacity has been installed since 2000 than the capacity of fossil and nuclear power. In 2016, 86% of all power capacity additions were of the renewable type.

Since 2008, renewable power has made up more than half of all new power capacity installed in the United States. In 2016, PV and wind power accounted for 60% of all new capacity installations. PV was number one ahead of wind and natural gas power.

Since the year 2013, China has been leading the world on renewable energy matters: on new wind power installations, PV, hydro, and solar thermal. In China, it is also PV that has recently become number one for new power installations.

All in all, in 2016 clean renewable power represented 55% of all new power capacity installations worldwide. Virtually all countries are concerned. As an example, in 2016 nine countries in North and South America, Asia, and Europe each had already over 10 GW of wind power installed. In 2015, the global capacity of wind power passed for the first time that of nuclear power. Following the International Energy Agency (IEA) in Paris, global renewable power capacity passed for the first time that of coal fired in 2015 as well: The world renewable capacity reached 1985 GW (31% of the total world power capacity) and coal power plants stood at 1951 GW.

"This is a whole new world".

Since the turn of the century, the renewables attracted over $3 trillion (3000 billion) of private capital investment. As PV and wind were newcomers on the markets, political support had to be expected. It is important to realise, however, that the conventional energies benefited and still benefit today from much higher support—and that one in cash. In Germany alone, national coal exploitation received since 1957 €200 billion in subsidy. Recently, the G20 heads of state decided to discontinue all energy subsidies by 2025. However, such declarations are not legally binding.

It is profitable to invest in the renewables. They are cheaper than the conventional fossil and nuclear energies. That explains their success in the world markets.

The various activities involved in the renewables exploitation, marketing, production, installation, operation and maintenance are rich in terms of job creation. Since the turn of the century, 10 million jobs have been created worldwide. In the United States, nowadays 260,000 people are working in the solar PV business compared to 50,000 in the coal business. And it is better for your health to work on clean solar panels than it is to work on dirty coal.

The massive introduction of solar energy and the renewables opens new perspectives for our lives. In Europe, the United States, Japan and Australia, well over 6 million families have gained some new energy autonomy with the recent installation of PV on their homes. This means better protection against the anonymous providers of the centralised conventional energies and their investment decision on production and distribution of energy we may disapprove. The renewables offer more transparency, freedom of decision and a sense of well-being when connected to clean energy instead of the dangerous and polluting conventional ones.

With PV, we are part of the modern semiconductor world. It goes even beyond the "silicon valley" with smartphones and the immediate communication via the Internet. It involves benefiting from important new satellite applications beyond GPS, communication and observation. It means a more convenient life at acceptable cost—living in more comfortable homes and sustainable city structures combining work and leisure in one place.

The world of bio-energy that we have to address later as well is a very important aspect. New perspectives of biomass production in agriculture and sustainable treatment of biological waste streams have to be considered. The new opportunities for the development of the poor in the "Third World Nations" are perhaps the most important aspects of solar energy that deserves its name.

What we want to address, too, are the more obvious relations we are going to have, as solar energy adepts, with nature, with our Sun, with the Universe.

Part 1

The Sun and Us

Chapter 1

The Legacy of the Sun

1.1 Man in the Universe

The Universe is quite a place. And we, the humans, don't count very much in it. Our role is at best that of an observer.

Until not so long ago, it was conventional wisdom that man and Earth were the centre of the Universe. Remember Nicolaus Copernicus, Kepler and Galilei, who were the first to show that the apparent evidence that the Sun and all stars turn around us is wrong, and that it is just Earth that rotates. Only "yesterday", in 1992, Galilei's honour for having stuck to the truth was fully rehabilitated by the Vatican.

Our Earth is indeed not the centre of the solar system, and the Sun with its planets is not sitting in the centre of our galaxy either. That one is occupied by a star-eating black hole. The Sun evolves inside a side arm of our galaxy that was formed 8.8 billion years ago. The Universe is 13.8 billion years old and the solar system is 4.6 billion years young. There is no sign that our galaxy had a particular central role in the Universe either.

Our Sun is an average-size star. Those stars have a lifetime of some 10 billion years and end up as "red giants". The larger a star is, the shorter is its lifetime. There are "massive stars" with 100 times the mass of our Sun; they live only a few million years

The Triumph of the Sun: The Energy of the New Century
Wolfgang Palz

ISBN 978-981-4800-06-8 (Hardcover), 978-0-429-48864-1 (eBook)
www.panstanford.com

before becoming a super giant star and exploding as a "supernova". They can be 500,000 times as luminous as the Sun. Supernovae in turn end up as neutron stars or black holes. Black holes can have a billion times the mass of our Sun.

Not only the masses and energies, but also the dimensions in the Universe are gigantic. Kilometres and miles being inappropriate to describe them, one uses the light year, the distance light travels in one year. Light is our messenger in the Universe. It is not weakened in empty space when travelling over 10 billion light years, a remarkable fact. However, as its speed is enormous but not infinite, we are unable to see the Universe as it is today. Only as it was millions or billions of years ago when the light departed from the objects we see now in our telescope.

The Universe is in a continuous movement following the laws of physics. There is factually no extra creation involved—unless you believe in one.

However, there are a lot of mysteries—one being the "big bang". How could it be that this enormous Universe developed at one point in time from a ball not bigger than a nailhead? Only in 1927–1929 Edwin Hubble and the Belgian priest Georges Lemaître discovered the big bang and the eternal expansion of the Universe. The expansion follows a precise speed, the Hubble constant.

Other mysteries concern the "dark matter" and the "dark energy" in the Universe. Only 5% of the Universe is visible to us, the rest is dark. The existence of the dark energy in the Universe is derived from the fact that the Universe is expanding forever. Otherwise the gravitational forces of the visible matter at stake should oblige the masses to stop expanding at one stage and restart coming back to the original point of the big bang. The other 27% of dark matter in the Universe is concluded from the observation that in spinning galaxies like ours, all stars orbit with the same speed around the centre, but they should not. Without the dark matter around, the stars at the edges of the spirals should travel more slowly.

Talking about mysteries, what can be the nature of the forces that keep the protons closely together in all atoms of the world? The protons packed in the nucleus of all atoms—except hydrogen—have all the same positive electric charge, and without a secretive force that keeps them together just in the nucleus, the atom should explode. The gravitational forces inside the nucleus are indeed infinite times smaller than the repulsive forces of the electric charges. In practice just when uranium or plutonium are bombarded with neutrons, some of the energies keeping the nucleus together are liberated, and as we know they can be terrible.

And at last, why do the physical constants dominating the Universe have exactly the size they have. They all look really like random numbers; who picked them? Take the three universal constants: *c*, the speed of light; *h*, Planck's constant, also connected to light; and *G*, the gravitational constant. Why is *c* exactly 299,792 km/s. Why isn't it 300,000 km/s? Why is the force describing after Newton the mutual attraction of masses proportional to 6.674×10^{-11} Nm/kg^2, or otherwise expressed, why was the apple falling on Newton's head entailing his interest in gravity, not falling faster or slower.

Next to the three universal constants, there are a few more physical constants, such as Coulomb for the electric charge. For all of them, the same question arises. If one of them had been different from what it is, would we have a different Universe? As far as one can know, today these unit numbers are eternal. Researchers tried to find the slightest evolution of them in time. There was none.

In conclusion, it is a matter of fact that the proceedings in the Universe are evolutionary, obeying the eternal laws of physics. The same is true as we shall see later for the biosphere. There are no traces of any creation except the laws of physics themselves, and some mysteries such as the big bang one that are impossible to explain rationally.

1.2 A Heaven of Stars, One Sun

All the stars you see in the night sky are suns except a few of our fellow planets, such as Venus. All of these stars belong to just one galaxy: our Milky Way. As we know today, the Milky Way consists of 100 billion stars, or 100,000 millions of them. But there are a lot more. Since we can rely on sophisticated telescopes, we know that the Cosmos contains 1000 billion galaxies. And each galaxy has approximately the same number of stars as our Milky Way. A good approximation for the total number of stars is infinite! Yet at the same time, these huge masses evolve in space, which is virtually empty. The apparent contradiction stems from the fact that the distances are so enormous as well.

While there is no reason why most of the stars don't have their planets, the number of planets similar to Earth could be equally important. But don't try to communicate with them, the distances are prohibitive.

Everything is in evolution. US astronomers have concluded in 2017 from simulations that galaxies can exchange material between them on a large scale: 50% of the material in our Milky Way might have indeed stemmed from other galaxies after having travelled some billions of years.

There is no doubt that the Sun system owes its existence to one or several supernovae. As mentioned before, supernovae are super heavy stars that explode and leave gases of the heavy elements that don't exist otherwise in the Universe. Our Earth, its biosphere, and we as well are made of the material stemming from supernovae explosions—with the exception of the lighter elements such as hydrogen. All the water in particular contains hydrogen, and that stems directly from the big bang.

Since 1987, astronomers are actually observing such a collapsing supernova, 163,000 light-years away in a neighbouring galaxy. It is the nearest supernova discovered yet. It has been studied for the past 30 years. At its peak, it radiated like 100 million Suns. Most of the light was caused by the decay of radioactive cobalt produced when the star shrunk catastrophically and then rebounded in thermonuclear fusion.

Scientists believe that our Sun may have formed from the remains of such supernovae.

The general theory about it holds that everything started with a giant molecular cloud, 65 light years across, like those that exist still today in our galaxy. Such molecular clouds can be, like everything in the Universe, enormous: 300,000 times the mass of the Sun. Clouds may form and dissociate in less than 10 million years. It is thought that the Sun formed from a "proto-planetary disk" within less than 50 million years—a relatively short period on the scale of the Universe. The Sun was not formed alone but in a cluster of between 1000 and 10,000 stars.

It is interesting to note that more recently, one came up with the idea that things were actually even a lot more complicated than that. And it is interesting enough to be reported as well in our book about the Sun. In 2012, the French specialist Mathieu Gounelle published his new modelling derived from work on asteroids, actually the big presence of magnesium 26 and nickel 60 in them. According to him, a big nebula collapsed 4.6 billion years ago entailing the creation of a first generation of as many as 5000 stars. Five million years later, the massive ones exploded as supernovae and ejected their elements. Again, 2 million years later, the left cloud collapsed, which led to the formation of a second generation of stars. Some of these stars were very massive with 30 times the mass of our Sun. Eventually one of those heavy stars ejected 100,000 years later the material that gave rise to a third generation of stars, among them our Sun and a hundred others. **This happened 4.5682 billion years ago—we shall see later when discussing the asteroids, how the Sun's exact age was found out.** The Sun's sisters disappeared into the galaxy. Some million years later, the massive star that had ejected the material to form the Sun and its sister stars passed away in another supernova. The Aztecs called her Coatlicue, the mother of the Sun and mother of the Gods.

Again, the Sun with its planets is now a bit less than 4.6 billion years old. It has not yet passed half of its life as it has yet 74% of hydrogen, its fuel, and 24% of helium on board. Eventually, for the Sun, a full age of 10.5 billion years is expected. As a result of small contraction of the core, the Sun's irradiance increases by 7% every billion of years; today it is 30% stronger than in its early life. Thermonuclear fusion takes place in the core at 15 million degrees Celsius or K. It takes some 100,000 years for the heat to reach the surface. That one is fortunately only at 5,778 K. If it were millions of degrees hot like the core, the virulent radiation would immediately burn away everything on Earth, including all the seawater. The "black body" radiation that is emitted by the Sun is consistent with the temperature on its surface and lies in the green-blue visible part of the spectrum. The best power intensity arriving on Earth before absorption effects by the air is 1,367 Watt per m^2. After absorption in the air, the light intensity comes down to 1 kW per m^2 on the ground at sea level.

Recently the solar radiation over Europe between the two periods 1965–1988 and 1989–2012 was compared to see the impact of pollution. It was found that irradiation increased in the latter period, by 2 to 3 Watt/m^2. It is not much but a measurable effect of reduced pollution. The acid rain that was common in the last century in Europe seems also to have disappeared.

The energy radiated by the Sun is of unimaginable magnitude. Every second, 627 million tonnes of hydrogen are burnt in the Sun's core into helium—that element rightly called after the Greek word for the Sun. An incredibly large amount. As the Sun radiates in all directions, all the energy Earth receives makes only less than a billionth of what the Sun emits at any time. In 2017, researchers discovered another surprise. According to the data collected by the Solar and Heliospheric Observatory (SOHO), the Sun on the latitude of its equator needs 25 days for one rotation. At the core inside, it is much different and closer to the rotational speed of Earth; it is one week for one revolution.

Europe has tried to imitate the Sun's performance. For 30 years or so Euratom has been tinkering with a thermonuclear project ITER in the south of France to generate commercial energy with a fusion reactor one day. Financing for fusion research—€30 billion so far—comes from public sources without much democratic control. Many think it is a waste of billions of Euros and one should leave the Sun where it is.

1.3 The Way the Sun Produces Its Energy

The Sun generates its energy by thermonuclear fusion in the core, where under enormous heat, four protons, the nuclei of hydrogen, come together to form a helium nucleus. In the process, two of the protons are converted to neutrons and additionally are created two positrons and two neutrinos. The neutrinos can be measured on Earth, though with much difficulty, to confirm the theoretical model.

The explanation sounds straightforward, but there is a major headache involved: The positive electric charge of the protons being the same for all creates a tremendous repulsion force between them. They don't hurt each other in any way and don't get in direct contact even at the high velocity they gain at high temperatures.

The question of how the Sun produces its energy always raised much interest among scientists. In the 19th century, Helmholtz and Lord Kelvin proposed gravitational contraction of the Sun. However, they found out soon that such energies would by far not be enough to explain the long existence of the Sun. When radioactivity was discovered in the early 20th century, it was proposed to be the source of the Sun's energy, but this theory also proved to be insufficient.

In 1920, **Arthur Eddington** in England rightly proposed the fusion of hydrogen into helium. It has been measured that the weight of a helium nucleus was somewhat lower than that of four hydrogen nuclei, and the difference in mass could explain, after the Einstein relation, all the energy produced. Eddington looked far ahead when he raised the question of "controlling this power for the well-being of the human race—or for its suicide".

Hence, Eddington was right but still the practical question of the Coulomb repulsion of the protons remained unsolved. Another breakthrough came in 1928 when the Russian-American **George Gamov** introduced quantum mechanics and the non-zero probability of two charged particles to overcome their mutual electrostatic repulsion. That was actually the right explanation, but the discussions between the protagonists of the time, Teller, Bethe, von Weizsäcker and many others went on. They knew each other well and discussed the subject for years.

In 1937, Carl Friedrich von Weizsäcker came up with what since came to be known as the **Bethe-Weizsäcker** cycle, or the CNO cycle. It uses as catalysts the elements carbon, nitrogen and oxygen (hence the abbreviation) that are also contained in the stars in small amounts to get the four protons into helium. Hans Bethe, the German-American who later played a big role in Los Alamos and the building of the first atomic bombs, got also involved in 1939 and proposed the same CNO cycle for the way our Sun generates its energy. On the other hand—actually at the same meeting in Washington DC—direct reaction between the protons had again been proposed by George Gamov and **Critchfield**. They were right and Bethe was wrong. Both processes, direct reaction of the protons or the CNO cycle, take place in the stars. However, in our Sun, the direct reaction prevails and the CNO cycle works only in heavier and hotter stars.

One can speculate that it is fortunate that the direct fusion of the protons is much obstructed by the enormous repulsion forces between them and that the rate of fusion is kept to a minimum by the quantum effect. Otherwise all protons would have combined at once and the Sun would have immediately exploded.

1.4 The Sun, Earth, and Us

1.4.1 Children of the Sun and Earth

It makes sense to call Earth our mother and the Sun our father. Indeed, in the Latin languages such as French and Spanish, one is female and the other male. However, it depends how you look at it. In German, both are female and the moon is male. Obviously, they got it wrong.

One is inclined to overlook in our everyday life that our existence is well regulated by the control the Sun–Earth couple has on us. Take sleeping. All life is linked to sleeping. No sleep means death. Sleep has its origin in the day and night movement of Earth in the light of the Sun.

Or take the jet lag. After travelling over longer distances, one has to adapt one's watch to the local position of the Sun that dictates the time there.

1.4.2 The Birth of Earth

Probably Earth and all planets originate from the same flat nebular disc that gave rise to the formation of the Sun. That must have happened shortly after the formation of our star, some 10 to 100 million years later. This view is consistent with the fact that all planets move in the same plane and on orbits the same way the Sun rotates.

The first to speculate that the planets were formed by condensation from a rotating nebula were **Kant** and later **Laplace**. In 1943 **C. F. von Weizsäcker**—who worked also, as we have seen, on the energy production inside the Sun, and the energy keeping the protons together in the atoms, too—came up with an extensive hypothesis on how our Earth and all planets may have formed. He explained why the planets' orbits have regular distances from the Sun following a precise sequence (the Titus–Bode equation). In particular, he assumed that the planets must have had originally the same composition of elements like the Sun, that is over 98% of hydrogen and helium and the small percentage left of heavier elements. On the planets closer to the Sun such as Earth, the light elements were ejected in the rotation process and only the heavier ones remained. As the outer planets further away from the Sun were cooler and icy, they retained the hydrogen. Indeed, the hydrogen content of Jupiter is proportionally comparable to that of the Sun.

Nowadays we know more precisely the exact chemical composition of Earth and the Sun. Earth contains mainly the four elements oxygen, iron, silicon and magnesium. All others don't make up more than 0.3% of Earth's total mass.

As mentioned before, all the elements were previously formed in massive stars at temperatures more than 100 times hotter than that in the core of our Sun. After the exhaustion of hydrogen, the stars start to burn helium. By contraction, the stars' temperature continues to rise. Carbon and oxygen are produced first, and then magnesium, followed by silicon and, after reaching 3 billion K, iron. For the stars, iron formation is the signal to explode in a supernova, and in this process all the elements heavier than iron are formed at last.

Interestingly, the Sun has always contained the same heavy elements as does Earth, but obviously only to the proportion of 1.76%—the rest being hydrogen and helium. The proportion of the heavy elements is not exactly the same like in Earth, but it is very close, with oxygen having the highest concentration like in Earth, and the others, iron, silicon and magnesium, being relatively frequent in the Sun as well.

Many more models about the planetary formation have been proposed until today. It is one of the subjects where nothing is final until the last physicist has made his comment. What prevails presently is the so-called SNDM model proposed by the Russian scientist Victor Safronov. It is not essentially different from von Weizsäcker's model.

1.4.3 Asteroids and Comets

Hence, the essential process of forming the bulk of our Earth was the collision of cosmic dust in an orbiting flat disc. That one was followed by the ejection of most of the hydrogen and the massive bombardment with asteroids and comets. Not to forget that primitive Earth was impacted by a smaller proto-planet with the creation of the moon.

The asteroids, those icy rocks that hurt Earth frequently stem from the "Kuiper belt", the space beyond the planet Neptune. They are remains of the formation of Jupiter and hence are as old as the solar system. As they orbit the Sun, they often can cross Earth's orbit.

It is thought that the comets, unlike the asteroids fly in from outside the solar system. There are billions of them and 184 have been identified for periodically orbiting the Sun. They are composed of ice, dust and gases in the centre. Their tails get 10 million km long. After some 500 orbits around the Sun, they become simple rocks, just like asteroids. A remarkable exploit was the visit of the comet "Tchouri" by a man-made satellite in 2016. By studying the isotopes of xenon in that comet and comparing it to that in Earth and the Sun, it was concluded that the kind of water in the comet was different from that in the solar system.

At the birth of our Earth, 1% of the water might have come from comets and only a 1/100,000 of its mass. The rest of our Earth's water may have come from asteroids and the rocks of the original disc that formed Earth in the first place. It is also thought that comets brought to Earth the first organic molecules, amino acids like glycine that stand for the origin of life.

The formation of our Earth must have been completed pretty quickly, creation of the moon included. Earth's oceans were already formed 4.4 billion years ago.

An exhibition at the Muséum National d'Histoire Naturelle in Paris in 2018 was devoted to the "Meteorites between Heaven and Earth", as it was called. Samples of the many different types of asteroids were on show. There are in particular the "survivors", asteroids that stem exactly from the time of the Sun's formation. Figure 1.1 shows a cross section of such a sample. There are bright grains enclosed in a brown matrix. The grains were formed at very high temperatures in the neighbourhood of the native Sun. From them, one could determine the exact age of our Sun as mentioned previously. The matrix stems from the dust out of which the Sun and its planets were originally formed—it is older than the solar system. Just as a reminder, the solar system formed from an enormous cloud containing over 98% hydrogen and 1.7% of dust, the baked dust we see on these asteroids, called chondrites by the experts. Our Earth originates from such dust, too.

Figure 1.1 View of an asteroid made of the original matter out of which our Solar system was formed (picture by the author).

Other types of meteorites are rich in carbon and organic material, the basic ingredients for life—but no traces of life as such.

Interestingly, most of the asteroids collected on Earth are of pure iron. Before man had learnt producing iron from iron ore—the start of the Age of Iron, people had no access to other sources of iron, for instance, to form weapons. It is claimed that even a dagger in the tomb of Toutankhamon was made from the iron of an asteroid.

The iron recovered from asteroids is the same as the iron in the heart of our Earth. Many asteroids were composed like planets of an iron core, the mantel, and a crust. Vesta is the name of a sample of this type that was recovered in bits and pieces. Such asteroids had even volcanism.

1.4.4 Earth, Ready for Life

Earth benefits from a few particular favours for life in its environment. First, there is the big Jupiter next to us, which prevents asteroids from impacting Earth on its orbit. Then, life was able to develop with all the main ingredients that were brought along when Earth was born, including carbon, which is the basis of all bio-chemical processes and water. All together, the elements of life are called CHONPS, which stands for carbon, hydrogen, oxygen, nitrogen, phosphorus and sulphur.

And then there is Earth's relation to the Sun. The molten iron alloys in its outer core are in continuous move and produce a dynamo effect. Without the magnetism created this way, the dangerous solar winds of protons would hurt Earth without hindrance and strip away from it the upper part of the atmosphere. To maintain the sequence of the seasons as they are, it is important that the axis of Earth rotation angle with respect to the plane of its rotation around the Sun remains stable. This is ensured with the help of the moon. On Mars, which lacks a moon, the inclination of the rotational axis may change from 10° to 70°.

The temperature is indeed very comfortable when considering that Earth is a dangerous place: The temperature in outer space is not higher than 20 K and Earth's surface where we live sits in the middle between the centre of Earth at 7,000 K and the Sun's surface at 5,778 K. The heat inside Earth stems from the time of its formation and the decay of uranium 238 with a half-life of 4.5 billion years. That heat manifests itself well in volcanism but plays not much of a role for the temperature on Earth that is dominated by the irradiance from the Sun. Thirty-one percent of the Sun's irradiation is reflected back into space—that makes the nice blue colour of Earth seen from outside. All the absorbed radiation ends up as heat. Also that is absorbed, for instance, to generate man-made electricity or mechanical power. Earth absorbs the Sun's irradiation in the visible spectrum at a wavelength of around 0.5 micron and rejects the same amount as heat in the infrared at some 10 microns. The resulting equilibrium temperature comes out as 254 K or −19°C on average. Fortunately, there are the greenhouse gases (GHG) that turn that icy cold temperature up 35°C into an enjoyable 15°C, Earth's average temperature.

We are going to see later how life actually developed on Earth. It was a winding road, most of the time catastrophic for the species that evolved, with five major mass extinctions. Organisms survived only thanks to their extraordinary vitality. Free oxygen in the air has existed for only over 2 billion years. However, life is already 3.7 billion years old; it was anaerobic initially.

1.4.5 The Last Ice Age

The last Ice Age came to an end only very "recently", some 9,000 to 10,000 years ago. It was a total cataclysm. And when it finished, it gave rise to a flood that was an even more terrible disaster.

The Ice Age lasted for 100,000 years. If one considers the many things that can occur in just 1,000 years—even the discovery of America happened only 500 years ago—one gets a weak impression of the enormous duration of 100,000 years for the population that lived at that time. Actually they were our direct ancestors. According to the latest discoveries, *Homo sapiens* has been around in all known territories already for some 300,000 years. Not only in East Africa, where the favourable climate is the best to preserve human remains, but also in Morocco, on the Balkan, or in Germany, traces of the first modern men have been found. Such a long time of 100,000 years for the Ice Age represents some 3000 generations of our ancestors. They had ample time to move around.

During the latest Ice Age—there have been several others before—32% of the landmasses were covered with ice, up to 3 km high. The high ice masses covered all the Northern parts of Europe, Asia and America, and the most Southern areas of the Southern hemisphere as well. Mountains like the Alps or Tibet were equally covered. Air enclosures in polar ice probes prove that the GHG in the air were reduced by half at that time. The cooling effect may have been further amplified by some major volcanic explosions. Eruption from Toba, a volcano, 74,000 years ago seems to have been one of them, obscuring the air and preventing the warming Sun radiation from getting through.

Living conditions must have been terrible. Probably not so many managed to survive—perhaps, in caves such as those in Southern France and Spain with wonderful paintings of that period or in caves in Germany where impressive artefacts were found recently. The few survivors were certainly at the origin of the language trees, the Indo-European languages branch being just one of them. English, German, French, Spanish, Russian or Indian languages are part of that one, but not the African idioms in particular. The splitting between the African and European populations must have occurred very early on. Despite the stress *Homo sapiens* suffered in that long period of ice, he proved to be very intelligent, not only in arts but also in creating our original languages. What we speak today are only degraded dialects from the one they invented.

1.4.6 Hephaistos

Eventually temperatures increased and the ice melted away in several steps. After some 9000 years, by 7000 BC, two-third of all ice had melted and that long-lasting Ice Age had finished. It is still not totally clear what phenomena led to the progressive warming again. First, one tried the natural variation of the Sun's irradiation by the so-called Milankovitch cycle. Earth's orbit around the Sun, which is slightly elliptical, indeed undergoes a precession because of a gravity effect from Jupiter and Saturn. Moreover, the tilt of Earth's axis varies between 22.1° and 24.5° over a cycle of 41,000 years. The solar irradiation received varies slightly with Earth's orbit evolution. The axis inclination had a maximum 10,700 years ago, while presently at 23.44° we are on a slight cooling trend. And there are several other cycles of Earth's move around the Sun. Since Plato, one knows that Earth passes every 2,150 years through a new sign of the 12 signs of the Zodiac. Presently we are at the age of Aquarius.

But it is considered that the climate change involved in the Ice Age was much too extreme to be interpreted by the slight changes of solar irradiation and temperature. Instead, the impacts of asteroids or comets have been proposed.

Some researchers think that a comet named Hephaistos must have played a particular role in what happened. That comet was one of the 400 centaurs that have been identified. With their 50 to 100 km size, centaurs are bigger than most other comets. Hephaistos must have been disintegrated when approaching an inner planet of the solar system some 100,000 years ago. It burst into million pieces with many of them hurting Earth. Many impacts have been reported towards the end of the last Ice Age. A first one 29,000 years ago and another 17,000 years ago leaving traces in the ice of Antarctica. In 1972, the Landsat satellite discovered in Alaska at Sythylemencat the crater of 500 m diameter asteroid that dated from 12,000 years ago. A 250 m diameter block came down in the Alps in Austria (Köfels) 8500 years ago. It is at the origin of the myth that heaven fell down.

The ancient Egyptian *Book of the Dead* 6000 years ago reports about a succession of cosmic catastrophes. Was the Sahara victim of a cosmic impact 7000 years ago provoking a local climate change? The Sahara became a full desert only some 7000 years ago. Over 3000 years ago, the Egyptians reported about a comet that passed from India to the North Sea in a fireball, with an earthquake and a tsunami that followed. The people of the sea in Egyptian history may have their origin in such impacts and also the legends of the 10 plagues of Egypt reported in the Bible. In living Chinese legends, the dragon Kong Kong destroyed one of the pillars of heaven with dramatic consequences. They go back to astronomical consequences 4,350 years ago.

1.4.7 The Great Flood

The consequence of that big ice melting was a flood of apocalyptic dimensions. It lasted for a few thousand years and brought torrential rains and gigantic waves with kilometre-high tsunamis. It is thought that most of our ancestors perished together with millions of species of animals.

The sea level increased by 130 m. What this means can be best understood considering that it rose by just a few centimetres owing to the climate change since industrialisation began. Everywhere the coastlines were moved inland. England became an island again and the land bridge between Alaska and Siberia went under water as well. The Mediterranean basin filled up and reconnected with the Atlantic ocean. As all the ice produced fresh water, the salinity of the seas was reduced, too, with the need for the fish to adapt.

The big flood was kept in memory by our ancestors. It is a tradition and was mentioned first in old Sumerian texts. The Bible has it also as the legend with a great flood and Noah the saviour of the biosphere.

1.4.8 The Paradise

Another legend of the Bible is the story of a paradise at the beginning of all times. Lately researchers think to have found it. They think it must have been there where nowadays a busy fleet of tankers carries much of the world oil, the Persian Gulf. Today the water level there is down to 90 m deep, 50 m on average. In the Ice Age before the flood, it was a land area with three rivers providing fresh water. It is speculated that man had an easy life there and the local climate was just right to provide food and a comfortable life without much effort.

As a matter of fact, the nearby Sahara on the same latitude, before becoming hostile to life and a desert, is known to have been populated towards the end of the Ice Age by nomadic tribes and big animals such as elephants and lions. Part of the Sahara is Egypt. It is speculated that an important civilisation came about there already in pre-pharaonic times when it was less arid. Much discussions turn around the sphinx in Giza. It was built as a lion and later its head was transformed into a pharao's head that is disproportionate in size—a sign of the later change. The body of the lion has all over the traces of heavy rainfall, a hint that it was built perhaps 7000 years ago, before the great flood.

1.4.9 The Cradle of Civilisation

The speculation of a paradise flooded later to become the Persian Gulf gets some support from the fact that the world's first civilisation started just a bit North of it, in Mesopotamia, what is now Iraq.

Human settlement can be traced back there 7000 years ago, in the lowest level of cities like Nineveh or Uruk. That was just after the flooding had come to an end. The Sumerian people, who were identified as the first to have lived there, invented it all: They established the first permanent settlements, built the first cities and kingdoms, and organised agriculture, irrigation, husbandry, administration and the rule of law. The Sumerians were among those who invented the first wheel. They invented pottery. They invented 200 years before the old Egyptians the first writing. They were the ones who first divided the circle into 360°. They invented "capitalism"; for the first time, one could possess personal wealth. They invented before the Bible the week of seven days.

A bit later, the Babylonians, also in Mesopotamia, put the year in 12 months at 30 days; they divided an hour into 60 minutes and a minute into 60 seconds. Till today the decimal system has remained unsuccessful in doing that.

The story of Gilgamesh, the hero of Sumerian Uruk, is the beginning of world literature.

1.4.10 Waiting Disasters

Today's mankind is threatened by three main dangers: meteorites, climate change and atomic war. The latter two will be discussed later.

To conclude on asteroids and meteorites, we may at last mention Toutatis, the asteroid called after a Gallo-Celtic god. It is part of the series called Apollo, which contains over 8000 asteroids, some of them as large as 10 km^2. They all regularly cross Earth's orbit. Toutatis comes near Earth every four years; the closest was in 2004, when it passed just four times the distance between Earth and the moon. Its size is impressive 4.6 km by 2.4 km. That's not too much smaller than the 10 km diameter one that led to the extinction of dinosaurs 65 million years ago.

We are familiar with Toutatis; a Chinese space probe succeeded in flying over it and taking pictures.

And there was also Florence. That asteroid of 4.4 km diameter—even a bit bigger than Toutatis—passed Earth just recently in September 2017.

Chapter 2

Energy for Life

2.1 What's Good Energy?

It goes without saying that energy is of fundamental importance for our existence. Imagine for a moment that all energy supply were suddenly stopped. Then life would come to a standstill. Energy for heating and cooling, energy for transport, and electricity drive our economy and determine our well-being and that of our environment. Even in agriculture, which is by and large mechanised, energy cut would have dramatic consequences and affect our food supply.

Total energy expenditures in the world amount to $1,800 billion a year, 2.4% of the GDP. Hundreds of millions of jobs are involved. Pollution and climate change are the consequences of unsustainable energy consumption. Along with the almost exponential growth of world population thanks to industrialisation, energy supply and consumption got the monumental dimensions they have today.

We are going to come back to all the different kinds of energy. The main emphasis has to be on electricity as it is the most sophisticated in use and the most complex to produce. Most of electricity is employed in engines, in industry, and for robotics and process control in manufacturing. It provides all our lighting,

The Triumph of the Sun: The Energy of the New Century
Wolfgang Palz

ISBN 978-981-4800-06-8 (Hardcover), 978-0-429-48864-1 (eBook)
www.panstanford.com

too, all the supply to meet our demand in informatics and consumer electronics.

Given the mega importance of energy, one should better get the inherent investments right. Here politics play a leading role and often those got it terribly wrong. Hundreds of billions of dollars have been wasted. Examples are in particular the world's nuclear programmes that were decided by over-optimistic politicians who launched major governmental programmes, financed from state budgets; and the blind investments in always new coal power plants without considering the alternatives.

What are eventually the credentials of a "good energy"?

The important elements that impact any energy option for investments on a national or regional scale are

- security of supply;
- compatibility with a clean natural environment and the world's climate;
- optimal cost in production and use, profitability of investments;
- decentralisation of supply, local development, creation of income, social comfort of working and living conditions, development of industry and craft;
- international co-operation and peace.

At stake are trillions of dollars and such important things as fighting mega-city development, preserving rural interests, fighting the always increasing divide between the rich and the poor, development of the less developed economies and the rural poor there, fighting ethnic and civil conflicts and wars and the misery of flight and displacement.

It sounds ambitious. But a reasonable energy policy can help.

2.2 The Good, the Bad, and the Ugly

2.2.1 The Good

Solar energy and all energies derived from the Sun's generous supply to us, bio-energy, hydro-electricity, and wind power can be counted in general terms as good energies. But they are all worth a closer look.

First, they belong to the carbon-free energies. Those get as such a tremendous support in the frame of the climate-change discussions that are so popular since the COP 21 (the Conference of the Parties—the countries) in Paris agreed in 2015 on a limitation of GHG emissions to preserve the climate. But be careful, the low- or zero- carbon energies carry with them a stowaway, the atomic power. One could notice at that big political conference all the nuclear lobbyists stood on the bridge to support a new opportunity they imagined through the climate change issue.

Let's be clear. Our climate cannot be preserved by all zero-carbon energies. It can only be preserved by the renewable energies! We come back later to the point that it cannot be nuclear.

A particular issue of controversy is **intermittency**. The meaning is that neither solar radiation nor wind flow is available continuously and the same holds for solar PV and wind power that employ them. But a fundamental demand for an energy supply is its availability anytime, over the day and during the year. Depending on the application, thermal or electrical storage devices can help to bridge that deficiency, at least for short intervals. As weather predictions become always more reliable, the availability of PV and wind becomes better predictable and so becomes planning to ensure the desired coverage of supply. When it comes to mainstream applications of energy, the holistic combination of the solar energies in their different forms is the obvious solution.

In reality, intermittency exists for all energies. It applies also for the conventional ones, for O&M, operation and maintenance questions. Take the figures provided by the World Energy Council for 2016. The average yearly operation time for bio-electricity was 4,500 hours, for hydro 3,700 hours, for wind electricity 2,000 hours and for PV 1,170 hours. All fossil and nuclear plants' operation time was far from continuous. At 4,000 hours, it was less than half the year.

And there is the split between solar PV that we are going to address later in much detail and **CSP, the concentrating solar power or "thermal solar power plants"**. The latter employ mirror concentration of the Sun's rays to produce first heat that is then converted into power. Compared to PV, it has several drawbacks. First, it comes only in large power units. "Conservatives" like that because they prefer big plants compared to the PV peanuts, the small decentralised ones. Well, today PV is the winner with over 400 GW installed worldwide and some 80 GW added every year, while CSP got just 4.7 GW in total in operation and not many new ones on the horizon.

Another disadvantage is the need for direct radiation that restricts its use to the "solar belts".

Its main drawback is that the cost of the electricity it produces is the double of that of PV. Eventually that was its killer.

But until today the technology has its supporters that appear often like a syndicate. Greenpeace counted among their supporters and the IEA in Paris. Google got involved but withdrew later. CSP has an association of its own, the European Solar Thermal Electricity Association.

Unlike PV, which involves thousands of manufacturers and installers, CSP has only a few of them. Abengoa in Spain was a global leader, praised in 2010 by Obama himself. No wonder that it got a loan guarantee of $2.9 billion from the US government. It has built 25% of all plants around the world so far. But from being an industry darling, it fell to become a financial invalid; in 2015, it lost $1.3 billion, and its stock market value was reduced to one-tenth. Business activities on CSP were in freefall.

Spain was a pioneer in CSP. Between 2010 and 2013, it installed 30 plants, with capacity between 50 and 200 MW each. By 2017, Spain had installed for 2.3 GW of CSP in total; it has not increased at all since 2013. In 2016, Spain's CSP plants provided a total of 5 TWh of electricity. This corresponds to a specific production of some 2.1 kWh per Watt and year comparable to the average production of all PV plants in the country. The CSP series' growth in Spain was cut after the country discontinued all support for solar power, for CSP and for PV alike.

Other countries that invested in CSP were Morocco, South Africa, Abu Dhabi and India (in Rajasthan). The Noor plants in Morocco are quite big. Noor I, II and III are completed since 2017. Together they have a nameplate capacity of 500 MW and cost €2 billion. Their electricity sales price of 19 cents/kWh is guaranteed by the government. It was totally financed by public money, the German KfW, the French AFD, the BEI and the African Development Bank. Construction of Noor IV has started, and it will partly employ PV.

The technologies in use so far around the world included parabolic trough concentration—with 90% of all plants the leader so far, tower power plants (solar furnace on a tower surrounded by mirrors), and Fresnel lenses.

Besides Spain, the United States was the greatest enthusiast for CSP. The United States operates by now a handful of plants, all with a capacity of over 250 MW each, in total for 1.74 GW. The world's second biggest CSP plant is IVANPAH in California. It consists of three towers with a nameplate capacity of 377 MW. It was built by Bright Source and Bechtel and cost $2.2 billion. It had a loan guarantee of $1.6 billion from the government. It opened in February 2014; later that year, Associated Press reported the plant produced only half of the expected output.

Behind the scenes, Germany was the great promoter of CSP. Groups at DLR, the nuclear centre in Jülich, and others, in cooperation with a solar centre at Almeria, Spain, were the long-time pioneers. As Germany does not have the climate to install such plants, one got the idea for "DESERTEC". That concept promoted the installation of large CSP plants in Northern Africa and transfer of the generated electricity to Germany. That was an absolute crazy idea from many points of view, but it got much support from some solar enthusiasts in Germany. The budget at stake put forward in 2009 was an unbelievable €400 billion. In the spirit of DESERTEC, a company was created in Germany, Solar Millennium AG. Twenty important companies became shareholders, among them Munich Re, Siemens, Deutsche Bank, RWE, and even the Club of Rome. With its connections in Spain, Solar Millennium was involved in the construction of many CSP plants there. The masterpiece was supposed to become a 1 GW CSP plant in Blythe, California. Eventually the truth prevailed: The stock market value of Solar Millennium lost 80% and the company declared bankruptcy in 2011. The official declaration was: PV is cheaper. The Blythe project was made smaller and the technology changed to PV.

With Solar Millennium DESERTEC died away as well. I was personally not unhappy about this outcome as I had fought together with Hermann Scheer in Germany to stop this nonsense.

The renewables have to handle opposition from a corner from where only support would have been anticipated, **the ecologists, the greens**. Often it is right to oppose the construction of new big hydro-dams. Is it also right to condemn wind power with the argument that it kills birds and degrades the landscape and living conditions in nearby settlements? Is it appropriate to put biomass on the sidelines? Worldwide bio-energy is the number 1 renewable energy in use. But many demonise it saying it is the source of hunger or unlimited deforestation. In Germany, biomass has no equal voice among the renewables. We come back later to all those misconceptions.

At the end, what makes solar energy and all the renewables a winner today is cost. **The renewables are the cheapest of all energies at stake.**

It is funny to notice that "experts" calculate already how much more it will cost to turn eventually all global energy supply to the renewables instead of sticking to the stinky conventional ones. That's all wrong. Not only renewable energies make this world cleaner and more pleasant to live in, but they make energies cheaper and more affordable.

2.2.2 The Bad

Coal is the dominating resource for electricity generation. It is the most polluting and leads the world's emissions of CO_2 gas. At its maximum of global utilization, 7.65 billion tonnes were burnt in one year, 2017 as it were. Over the last few years, global demand was relatively stable. Coal is the major contributor of the global emissions of the gigantic amount of 32.5 billion tonnes of CO_2 in 2017.

Of the total coal demand of 7.65 billion tonnes, 70%, i.e. 5.41 billion tonnes, went for electricity generation in 2017, slightly less than in 2014, when it had a maximum of use for that purpose. Further in this book, we shall concentrate on this part of consumption. The remaining part of global coal consumption is taken by steel making and other industrial and domestic uses for steam generation or heating.

The real cost of electricity from coal depends on the price that has to be paid to compensate for those CO_2 emissions and that one is a lot higher than the dumped price it has on the markets. There should be a CO_2 tax, but there is none. Instead CO_2

is priced in emission trades. The EU price there stands now at €6 per tonne. In reality it is considered that it should rather stand between $40 and $80 per tonne to compensate for the damage it causes and to keep climate change at bay. At the meeting of the G20 heads of State in 2017 in Hamburg, even a minimum price of $190 per tonne was demanded. China, the biggest polluter, was setting up in 2017 a national Carbon Trading Scheme to fight emissions. It is the biggest worldwide.

Not enough that a fair price to electricity from coal is not applied, it is further subsidised. As part of the fossil energies, it benefits from the $444 billion received as subsidies every year in the G20 countries.

2.2.3 The Ugly

Nuclear power is the subject of much concern. Most of the 441 nuclear plants connected to a grid globally have passed half of their life. Time gets nearer when they have to be dismantled. A gigantic financial and environmental mortgage that is. It costs as much to take down a plant than it had cost to build it and a mountain of nuclear wastes to dispose of. And the costs can also be a lot higher: The dismantling of the Fukushima plant is anticipated to cost €90 billion and it will take 40 years to finish. Who is going to pay for it? The taxpayer. There is no insurance for nuclear power plants available. The damage is unpredictable and can be gigantic; no insurance company takes the risk.

In October 2017 Greenpeace handed to the French government a detailed report about the risk of terrorist attack on the country's nuclear park. The conclusion was that the risks are high. It is urgent to strengthen the cooling pools and increasing security of the 58 plants in operation. It would cost €140–222 billion, three to five times of what the operator EDF has planned.

Building new nuclear power plants turns out to be a nightmare, too. In Europe, this century saw the start of only two new ones, one in Finland and the other in France (Flamanville in Normandy). The building cost tripled for both to some €10,000 per kW and both needed over 10 years to be built with many years of delay. Perhaps they start operation in 2019.

Seeing such a tremendous "success", the United Kingdom did not want to stand on the sideline and decided to let the French build two new reactors of French design in their country at Hinkley Point. The 3.2 GW foreseen would cost €24 billion. The builder is EDF, which carries a debt of €37.4 billion. That gives not much insurance about the practicality of the project, even if the French government gives its guarantee. The finance director of EDF thought nothing good about the deal and resigned.

Figure 2.1 The nuclear power plant at Tricastin in France (picture by the author).

The United Kingdom guarantees to EDF and the Chinese partner CGN a return price of 10.5 cents/kWh for 35 years. Very impressive. Yet the current wholesale price of electricity in the United Kingdom is 3.5 cents/kWh and the kWh price for wind electricity only a fraction of what nuclear could ever deliver.

2.3 Pollution and Climate Change

2.3.1 Pollution

Following some estimates, 19% of the world's arable land is contaminated with heavy metals and other pollutants. More than half of all groundwater is polluted with substances that are dangerous for man's health.

The World Health Organization (WHO) claims that 92% of the world's population lives in places where air quality is bad. It attributes 3 million deaths per year to the exposure to outdoor air pollution. In Europe, the EU Commission estimates the number of early deaths because of pollution to be 400,000 per annum. In 2017, it has put in place upper limits of emissions from lignite burning power plants. Germany was not happy and considers retrofitting of plants too expensive. Euracoal, the European lobby group of the coal industry, complained that by 2021 after the deadline of compliance, 4 out of 5 coal power plants in Europe will not meet the new standards. Billions of Euros of investments are needed, or some plants must be closed. Excellent news!

In China and India, dust, haze, and smog strongly affect even the visibility, in particular in the big cities. Coal consumption is a main contributor to this disaster. That is why China, which relies heavily on coal, has put in place a vigorous policy for change. One result is China's new leadership on clean renewable energies that it deploys massively in the country.

In almost all big cities in Germany, the NO_x and particulate matter exceed by twice the accepted limits.

The "Union of Concerned Scientists" in the United States analysed the pollution of its coal power plants and the resulting smog, acid rain, and toxic elements in the air. Most US plants have not installed pollution control of the flue gases they emit. Next to all that CO_2 blown into the air, a typical coal power plant of 600 MW emits thousands of tonnes of SO_2 and NO_x per year. And the country has some 600 of those polluting coal plants. They emit soot, fly ashes, lead, cadmium, toxic carbon monoxide CO, and even traces of plutonium. The most dangerous of them all is mercury, as it comes out to be 170 pounds per plant and year, and only 9% of US coal plants have devices in place to reduce its emission.

And it is more than a curiosity that several tonnes of plutonium is still in the air from the weapons testing in the 1950s and 1960s. The pollution arising from the nuclear power business is all too often swept under the carpet. When in France effluents of americium and plutonium are discovered near the nuclear processing plant at La Hague, the typical comment is that it is not dangerous to health. Only once it happened in France that Areva was officially condemned in court for pollution. It was at the occasion of important leaks of uranium at the Tricastin plant in 2008.

When the nuclear industry did not know what to do with all the waste it produces worldwide, it did not hesitate to go for ocean dumping of the waste. That started right away in 1946 and 13 countries participated, including the United States, Germany, the United Kingdom, Russia, and Japan. Over 150,000 steel containers were disposed of in the seas worldwide. And in the meantime, the barrels leak. *The Wall Street Journal*, which is not the voice of Greenpeace, reported at one stage that the sea floor 50 miles from San Francisco was polluted with plutonium 1,000 times above normal. By 1993, the dumping of radioactive waste in the sea was officially stopped by the treaty of London. The treaty runs until now in 2018. And Japan is already in the starting blocks to dump amounts of waste never seen before: 920,000 tonnes of Fukushima waste is intended to go into the sea. You like fish? Enjoy it.

The effluents emitted from the operating nuclear power plants are responsible for leukaemia and thyroid cancers. The nuclear industry and its supporters were by and large able to keep public opinion at bay about it: "Move along, there is nothing to see here". Many studies involving the medical profession made it clear that everybody can be concerned; there is no risk-free dose of radiation. Young children and pregnant women are the first to be at risk of getting leukaemia. It is a small risk, but a real risk.

2.3.2 Climate Change

Awareness of the dangers of climate change started, in particular, in 1987 with a report by Gro Harlem Brundtland: **"Our Common Future, From One Earth to One World, a Call for Action"**. The report came from the World Commission on Environment and Development, on behalf of the UN General Assembly.

The famous UN Earth Summit followed in 1992 in Rio de Janeiro, where 172 governments were represented. The conference agreed on a Climate Change Convention that led later to the "Kyoto Protocol". Alternative sources of energy to replace the use of fossil sources were addressed in particular. The Kyoto Protocol was signed in 1997. It had mandatory targets on GHG emissions but no mechanism to control implementation.

Also, after Rio was set up the Conferences of the Parties (COP) of the UN member countries. COP 1 was held in 1995 in Bonn. Since then, a conference has been held every year. COP 21 was held in Paris in December 2015 and COP 23 was again held in Bonn in 2017. The Paris meeting was a big global mobilisation. It was attended by all the important heads of state of the world. A treaty was agreed and signed by the participating countries. The United States signed it as well before taking the exit later when a new president had the say. Most states have also ratified it since then into law in their countries. The treaty refrains from mentioning coal energy as the bad one, the climate killer. It does not mention solar energy and the renewables either—all that to keep the countries known for their sympathies for the fossil energies in the boat.

In 2017, 13 US federal agencies, led by the National Academy of Sciences, came out with a report in much support of the climate change debate. Under the motto "Temperatures increased drastically in the United States since 1880", they insist on the fact that thousands of studies prepared by ten thousands of international scientists came to the same conclusion: The world's average surface air temperature has warmed by 0.9°C since 1880. Between 1951 and 2010, the human contribution to temperature increase was 0.65°C. The agencies project that even with low GHG emissions, by the late century, temperatures in the United States would increase by 2.8°C and with higher emissions by 4.8°C. The report came as the Fourth National Climate Assessment. It was leaked in the summer of 2017 to the press waiting for endorsement by the White House. Against the fact that the president had previously quit the "Paris agreement," eventually the White House approved its publication, without alterations.

What mobilised in particular the stakeholders in energy was the demand to limit temperature increase in the long-term future to less than 2°C, even though it means to some experts an over-ambitious target. It is not mandatory either. Fighting climate change has also become a grassroots movement that was strengthened by the Paris agreement. When Washington left the treaty for reasons of cost, California took up the lead for the United States. The biggest companies in the United States and the world signed declarations of support: The "We Mean Business" alliance subscribes to the objective to replace all fossil energies with the renewables by 2050. That alliance groups 490 enterprises worth $8,100 billion and 183 investment funds worth $20,000 billion. Major cities of the world declared their support, too. The "Under2 MOU" set up in 2015 by California and Baden-Württemberg in Germany is grouping 165 provinces in 33 countries.

The whole process initiated by the UN is scientifically accompanied by the International Panel on Climate Change (IPCC). It makes the calculations about what emissions lead to what temperature increase. The panel was established in 1988 by the United Nations Environment Programme (UNEP) and the World Meteorological Organization (WMO) and endorsed by the UN General Assembly. When it was young, the panel made a lot of noise in the media but has become more discrete in the meantime.

At one stage at last, solar energy and the renewables attracted the particular attention they deserve in all this debate about climate change. This happened with the "World Summit on Sustainable Development" in Johannesburg, South Africa, in 2002. At that meeting, the German chancellor invited the world to a specific conference on solar energy—well he did so since his party fellow Hermann Scheer had written it into the text of his speech. Then in 2004, took indeed place in Bonn, the German capital at the time, the first World International Renewable Energy Conference (WIREC). It was a big festive meeting, attended by 3,000 delegates from 154 countries around the world. The Declaration of the Conference was calling for the renewables to play a major role in the economy of the 21st century. Rightly so. And it was done! That's what we have to report about in this book.

Other WIREC meetings followed.

The 2005 conference was held in Beijing in the Great Hall of the People.

The big WIREC in 2008 was held in Washington DC. It took place on invitation of the Department of State; most other departments of the US administration attended, too. The US president gave a speech. The motto of the conference was energy independence through renewable energy. The general organiser was the American Council on Renewable Energy (ACORE) under its president, my friend Michael Eckhart.

Now that the dust has settled a bit, one can come back to the fundamentals. How dangerous is that climate change. It is a fact that the world's weather is more and more unpredictable with torrential rains, unusually strong heat waves, and big droughts. But the water level of the seas has not yet increased by more than 20 cm since industrialisation began. At the 3 mm added in 2016, one noticed that half of the level increase comes simply from the heating up of the seawater and the other half stems from the melting of the Greenland and Antarctica ice. Nothing to be alarmed about so far.

In the movie *The Day after Tomorrow*, from 2004, the question is raised whether the Gulf Stream that heats us so well in Europe could be stopped eventually by climate change. That would indeed be a major catastrophe. But models so far show that it will remain stable.

Another concern about possible feedbacks with the acceleration effect are the tundra in Siberia and permafrost in general. Actually a quarter of the landmass north of the equator is permanently frozen. By now temperature has increased by some 2 to 3°C there, and it is anticipated that by 2080 the frozen area could be reduced by a third. The GHG methane and CO_2 might be liberated this way and strengthen climate change further. So far this is not the case.

Here is other good news. The agreement in 2016 on the amended "Montreal Protocol" to protect the stratospheric ozone layer by phasing out the HFC in industry helps the climate also. The GHG effect of those HFC is indeed 14,000 times stronger than that of CO_2. And climate change may also have its good sides. Simulations seem to have shown that the Sahel region, which covers a huge area in Africa and is hostile to life because of an extensive drought, might now benefit from more rainfall that could double or even triple. And reality meets modelling: In the summer of 2017, Niger, one of the world's driest countries, had to fight enormous rain floods.

Another effect gives the lie to the simulation established on climate change: While in the 15 years from 1998 to 2012, GHG emissions increased by a third, global warming did not. All the better.

Whatever the simulation models, GHG emissions may simply not increase at all because the energy policy giving priority to those energies we receive from the Sun becomes effective a lot earlier than originally thought. Namely, thanks to China's drive away from coal, world CO_2 emissions stopped increasing already since 2013.

2.4 How Industrialisation Marginalised Solar Energy

2.4.1 The Traditional Renewable Energies

Before industrialisation started in the 19th century, the world relied almost exclusively on solar energy in its different forms. Wood burning and "Horse Power" or muscle power were the leading ones. Until WWI, the big cities like Paris were crowded with horses for transport and suffered tremendous pollution that this entailed. Wind and water wheels were used for water pumping and milling of grain. They were popular throughout Europe, the United States and China. Six million wind-powered water pumps were in use in the United States in the 1880s and thousands of them in Denmark for pumping and milling. Wind power had also a leading position for sea transport. Most ships were sailing vessels.

The first steam engines were fired with wood and the first gas engines with alcohol one had to buy from the pharmacy. But that did not last for long. The first to make massive inroads into the global energy markets was coal.

Modern forms of solar energy as we use them today were already well known but left aside—too sophisticated for the time. Solar PV was known as a laboratory item and Werner von Siemens, one of the fathers of industrialisation in Germany, saw its potential coming. And as coal had already been recognised as a polluter of the world, Augustin Mouchot started in the 1860s his solar activities with the objective to replace coal. He can be considered the father of CSP, too. Already in 1866, he was able to show to Napoleon III an engine to produce mechanical power—electricity as an energy carrier came in general use only in 1882—that employed light concentration with mirrors to produce steam. At the Universal Exhibition of 1878 in Paris, his solar printing machine was a colossal success.

The first electricity-generating wind turbines were invented in the United States and Denmark at the end of the 19th century, but they needed more time for development. Wind power will be covered in detail later.

2.4.2 The Victory Road of Coal

Coal for iron and steel making, steam generation for mechanical power in the new industries and the locomotives of the fast developing railroads benefited from an ever faster accelerating growth of its use. By the end of the 19th century, the world coal consumption had already increased from virtually nothing to 500 million tonnes a year. Coal was the first of the fossil fuels to take off. In the new electricity generation business that came forward at the end of the 19th century, coal had found an additional position of choice for firing all the new power plants.

It took 40 more years for coal use to double to 1 billion tonnes a year by 1940. Forty years later in 1980, it had more than tripled to 3.7 billion tonnes a year. Eventually by the year 2000, at 5.4 billion tonnes, it became just over tenfold the quantity used in 1900. In the meantime, oil and natural gas massively conquered the world markets, too, and coal made out no more than a third of the total of fossil energies in 2000. Curiously, the modern solar energies PV and wind power counted for just 0.3 thousandth of all energy the world consumed that year.

Since 2000, the world coal consumption has increased again by over 50%. In reality, all that increase goes on the account of China. China is a latecomer as a world's great economic power and it had to catch up. From 2000 to 2013, China almost tripled its coal consumption.

In the rest of the world, coal consumption was stable or rather decreased in this century.

The world coal power capacity generated 40% of the world's electricity in 2016. But since 2000, coal consumption has decreased by 10% in Europe and even 20% in the United States.

In Germany, it increased slightly. And correspondingly, Germany's CO_2 emissions did not decrease either. In 2017, Germany still produced 40% of its electricity from imported hard coal and lignite from open cast mines in the country. And it blows each year 900 million tonnes of CO_2 in the air.

India increased its coal consumption in the last decade by 6% a year; so it doubled there between 2000 and 2013. But consumption is not of the same order as China's, which is 10 times higher.

In 2013, half of the world's coal consumption was attributed to China; Europe came in with just over 10% of the global total—of which 20% for Germany—and the United States with just 10% as well. India's coal consumption equals by now that of the United States and of Europe.

The boom of coal consumption has come to an end, even in China and India. In a "Boom and Bust" report of 2017, the Sierra Club and Greenpeace announced a 50% decline in planned new coal power globally and a 62% decrease of new construction starts. China's government put on hold 300 GW of planned new coal power plants, including 55 GW under construction. Sixty-five percent of electricity in China in 2016 was generated from coal, 11% less than in 2010. Also, since 2010, the operational time of China's coal power plants has decreased by 20%. All together, China's coal consumption declined 4.7% in 2016 compared with the preceding year. As mentioned earlier, China's CO_2 emissions have stabilised since 2013.

In 2016, the United States produced 30% of its electricity from coal. In 1988, it had been 57%. US coal-related CO_2 emissions fell in 2016 almost by half compared to its maximum in 2008. Two major coal companies in the country are bankrupt: Peabody and Arch Coal; the rest lost most of their share price. One-third of US coal power capacity consists of old, dirty, and underutilised units. They are ripe for retirement and no more competitive with solar.

In India, the largest coal mining company plans decommissioning part of their coal mines considering that they are no more competitive. In May 2017, 14 GW of planned new coal power stations were cancelled. Solar is cheaper than coal in India. In 2016/17, 6.9 GW of new coal power was installed compared to 14.1 GW of renewable capacity.

In Europe, the powerful utilities' association Eurelectric has decided in March 2017 to deliver in a pledge to the COP 21 treaty and outlaw the developments of new coal power plants in Europe. Germany, which still operates 145 coal plants in the country, was reluctant to subscribe.

2.4.3 Nuclear Power

The nuclear age was declared in 1960 when the first GW of atomic power had been installed. Thirty years later, 300 GW of nuclear power was in place; many others had been already cancelled since 1970. With the exception of a few ones in China, no new nuclear plants became grid connected in this new century.

The nuclear age came to an early end. The final nails in its coffin were "Three Mile Island" in 1979, Chernobyl in 1986, and Fukushima in 2011. In 2016, 441 reactors were still on the grids. But many had stopped operation most of the time if not forever.

Japan originally had 54 atomic reactors in operation. By 2018, only 5 of them kept working and 14 had been definitely closed for reasons of security and excessive operational cost.

The year 1993 was nuclear's best year, when it produced 17% of global electricity. The share has decreased ever since.

By now, since the mid of the second decade in this century, the world's nuclear industry is in jeopardy. Westinghouse Electric Company in the United States builds and operates approximately half of the world's nuclear power plants and it is in deep trouble. Toshiba in Japan is its majority shareholder since 2006. Toshiba bought yet another nuclear company in the United States, Stone & Webster, in 2015 and that was one nuclear company too many. Cost overruns and delays in plants in Georgia and North Carolina led to the abyss. At the end of fiscal year in March 2017, Toshiba had to publish a loss of €7.5 billion in its nuclear business. To stay afloat on the stock market it has to sell its jewel, the memory chip business that has also Apple as a client. Toshiba is reconsidering its nuclear business.

The situation with Areva in France is not any better. Up to 2016, Areva, the nuclear construction and mining company, accumulated a deficit of €10 billion. Eventually the French government had to save it from bankruptcy. In an arrangement that was lately approved by the EU authorities in Brussels, EDF, the French power utility, took over Areva with two Japanese investors as minority stakeholders. The possible involvement of a Chinese nuclear company was expected but did not materialise so far. As the French government, with 83% stake, has the major control of EDF, the nuclear option is now by and large in the hands of the state. However, the deal did not go down well with the stock market. EDF's market value lost 10% in one day; it is now worth only a tenth of what is was 10 years ago.

On the ground, the affairs are an image of the institutional situation: The new nuclear flagship of France, the EPR at Flamanville, is expected to start operation end 2018. But there is a funny condition imposed by the French nuclear authority. As one is not sure about the reliability of an essential component of the plant, the operation has to be stopped again only after a few years to replace the risky part with a new one.

To the many accidents come tremendous risks and dangers. Take plutonium. Plutonium is the most toxic element; it does not exist in nature, it is man-made. All the reactors in operation produce next to the electricity and a myriad of wastes 116 tonnes of plutonium every year. Enough to produce 23,200 atomic bombs. Every year. Don't tell the terrorists.

In countries like France that rely heavily on atomic power for their energy supply, public opinion is split. There are those who see in nuclear, as long as no plant exploded yet in their country, a reliable source and security of supply. And then there are those who know better but are not sure how to get rid of all those plants. Future will tell.

The success of solar energy is no question for the future only. Since the turn of the century, the wheel has been turned back again. What the conventional energies took away the century before is taken back by the energy Earth belongs to, the Sun's.

Part 2

The New Century Is Solar

Chapter 3

The Solar Revolution of the Year 2000

Since the time energy supply became a mass market, there have been many reasons in support of solar energy development and use.

One of the first was the concern that the fossil resources in the ground could be depleted very soon. However, now that precise analysis has become available, one knows that the deposits of hard coal and lignite are enough for more centuries of use. For oil there had been more concern, but that was not counting with the barrels of oil from oil shale and tar sand in North America that came unexpectedly and massively on the world markets in the meantime. The resources of natural gas were long considered even tighter. This concern has also evaporated by now. Even uranium minerals are available at low cost. The contrary was expected some years ago by Areva in France; the wrong speculation on its price increase contributed to its bankruptcy.

A political concern of highest priority was always the criteria of national energy independence. The oil price crises of the 1970s and 1980s triggered by the conflicts in the Middle East were the examples still in our memory. What followed were the massive investments in nuclear power and coal.

The concern about climate change was always around but had not the priority it has today. In Western Europe, with its long

The Triumph of the Sun: The Energy of the New Century
Wolfgang Palz

ISBN 978-981-4800-06-8 (Hardcover), 978-0-429-48864-1 (eBook)
www.panstanford.com

coastline and the Netherlands already partly under sea level, the main interest of the general public was the threat of a major sea level rise.

The cost prospect of solar energy was a non-negligible incentive for PV development. Already since the 1960s, scientific evidence had it that PV could eventually beat nuclear in terms of cost: Nuclear cost was the benchmark then.

As a matter of fact, against this background of energy strategies for the future, the prevailing one that led to a change of paradigm in favour of solar energy, the solar revolution at the turn of the century was the pressing threat of a nuclear conflict, in the midst of the cold war.

3.1 The Threat of a Nuclear War

Currently, nine countries possess in total an arsenal of some 15,000 nuclear weapons, each of them many times more powerful than those dropped over Japan in 1945. Russia has 7,000 nuclear warheads and the United States 6,800. China has 270 of them and comes like the other nuclear powers far behind the two superpowers.

Germany has no bombs on its own but hosts 60 US tactical nuclear bombs stored at Ramstein.

At the hot phase of the "cold war" in the 1970s and 1980s up to 1990, the total arsenal was with some 70,000 warheads, almost five times bigger than today.

Currently, the trend goes again towards increasing the arsenal. The new US president made announcements in this sense. The United States estimates that $399 billion are needed to modernise its nuclear arsenal by 2026, Russia speaks of $330 billion needed for its nuclear armament over the next 20 years.

Since the 1970s, a lot has been done by the two superpowers to arrange some deals for disarmament. SALT I and II (Strategic Arms Limitation Talks) limited the number of bombs allowed on both sides and the number of submarines, bombers, and intercontinental ballistic missiles, too. An Anti Ballistic Missile Treaty ABM was signed, but in 2001 the United States pulled out of it.

A Strategic Arms Reduction Treaty (START) was signed in 1991. In 2010, a new START treaty was signed committing both sides to reduce the number of nuclear weapons to 1,550 by 2018.

Besides, an agreement was signed in 2000 to get rid of 34 tonnes of weapon-grade plutonium. It was renewed in 2009. However, in October 2016, Putin officially withdrew from it. In the meantime, the relations between the United States and Russia degraded much and continue to degrade.

Observers notice that now the threat of a nuclear war between the superpowers is higher than it was during the cold war. At the World Economic Forum in Davos in 2017, international business leaders declared that the possible threat of arms of mass destruction is the biggest element of insecurity of our time.

Robert McNamara published in 1987 his book *Blundering into Disaster*. When he was US Secretary of Defense, he participated in three world crises, Berlin 1961, Cuba 1962, Mid-East war 1967. Each had the potential to go nuclear, he wrote. Neither of the superpowers wanted a military conflict, but "lack of information, misinformation, and misjudgement led to confrontation". The Cuban missile crisis was closest the world has come to a nuclear war. The Russians had nuclear weapons in Cuba and came close to giving permission for their use against an American invasion, he continues. At one point, a Russian submarine had been commanded to surface by the US counterpart. The Americans did not know that the submarine carried a nuclear torpedo. It had already been armed by the captain. Only because the submarine brigade commander was on board, the torpedo was not fired as he overruled the captain. "So was diffused the threat of a nuclear attack on the American fleet".

In 2007, former Secretary of Defense William Perry has quoted the odds of a nuclear terrorist attack within one decade being 50–50.

Thus far, only the balance of nuclear terror between the superpowers has preserved the world from nuclear Armageddon. It is not desirable that there is only one superpower. The risk of a nuclear strike would be real then. The temptation was always there. In 1952, general McArthur advised President Eisenhower to use atomic bombs to end the Korean war. John von Neumann, a pioneer of atomic bomb development and informatics, wanted to exterminate the Soviet Union with nuclear bombs. President Nixon, among other qualities a drug addict, was in favour of using nuclear weaponry.

Nuclear warheads are all the time around us. They permanently circulate the seas of the world in submarines and the air above us in strategic bombers or wait in numerous missile launch pads. And there have been accidents. In 1950, a plane perished in British Colombia, Canada. It had a uranium bomb on board that was never recovered. On 17 January 1966, a B52 crashed at Palomares in Spain. It had four bombs on board. Two of them ruptured and dispersed 3 kg of plutonium kilometres around. A third bomb went into the sea. Twenty-eight naval ships searched for it for 80 days. They recovered it from 870 m under the sea. Almost 50 years later in 2015, the United States signed an agreement with Spain to definitely clean up the site; 1700 tonnes of contaminated earth had to be shipped to Carolina. A similar story happened at Thule Air Base in Greenland in 1968. Four bombs were ejected from a crashing plane. It took 9 months to clean the site. Thousands of cubic metres of waste was collected. It appears that one of the bombs was never found.

Eventually in July 2017, the UN adopted a new treaty prohibiting all nuclear weapons; 141 nations approved the treaty, but all the 9 "nuclear powers" voted against it.

A world without arms? As it stands, there is no chance this is going to happen. Every year, some $1,700 billion are spent anew in support of the military sector—$1.7 trillion or $1,700,000 million! In 2018, the United States, as always, was number one with $700 billion, an increase of 15% over 2016. Europe is second with $300 billion, including Germany with $41 billion. Then follow China with $215 billion, Russia with $69 billion, and Saudi Arabia with $63 billion.

The United States is the biggest weapon exporter at 33% of the world market.

The United States has also impressive intelligence programmes. At some $80 billion, they come higher than the whole military budget of Russia. A lot of this budget goes to "defence contracting companies", or investments like the large computer at the NSA.

3.2 A Society in Doubt about Its Future

Already in 1950, the World Peace Movement issued the Stockholm Appeal, calling for an absolute ban on nuclear weapons. It was signed by F. Joliot, Picasso, Th. Mann, and 500 million people throughout the world.

In Germany, in 1957, 18 nuclear physicists published the "Göttinger Manifest". It was signed by some leading scientists in nuclear physics in Germany and the world, including Hahn, Heisenberg, Born, C. F. von Weizsäcker, and K. Wirtz. We come back to these important personalities later in this book. On a personal note, Wirtz, who had participated with others at the German nuclear studies during WWII, was my professor later when I studied nuclear reactors in Karlsruhe. The purpose of the manifesto was to strongly oppose that the new German Bundeswehr be armed with nuclear weapons. And it was successful. After some hesitation, the German chancellor agreed. Shortly after, in March 1958, the socialist party SPD issued the motto, "Fight nuclear death".

Another manifesto that was successful, even though a lot later, was in November 2016, a call of 15 Nobel Prize winners on the UN to adopt a resolution against nuclear arms. It said that the dangers of nuclear arms are utterly unacceptable, and the only way to prevent an unthinkable catastrophe is to eliminate them completely. The UN resolution mentioned above is also a result of this call.

The biggest German demonstrations of all times took place against the stationing of new nuclear ballistic missiles in the country, and against a third world war that was seen as probable. After a first demonstration in 1981 in Bonn, the German capital, an even bigger one in 1983 brought together over 1.3 million people there. At the Bonner Hofgarten took place the main meeting with half a million people; 200,000 people formed a human chain over 100 km long in the south of the country.

In the early 1980s, a protagonist of the German peace movement was Petra Kelly, also called the "Jeanne d'Arc of the Nuclear Age". She was an inspiration for the anti-nuclear movement in Germany. The Bertrand Russel Campaign for a Nuclear-Free Europe was one of her many initiatives.

There have actually been many pioneers against the atomic threat and for a better world. Frédéric Joliot, a Nobel Prize winner of 1936, declared in 1952 "...we got to look very seriously and right away into the potential of solar energy utilisation". He was actually the father of the French nuclear programme, and that gives this statement particular weight. He was the first after Hahn's discovery of the splitting of a uranium atom to prove experimentally that this entails a chain reaction. He filed a patent on the atomic bomb.

Sergeï Sakharov is perhaps the greatest anti-nuclear hero the world has seen so far. The European Parliament regularly gives a prize in his honour. Originally Sakharov was the chief of the Soviet nuclear arms programme. On his account goes the Tsar bomb, the biggest bomb that has ever exploded on Earth. When he saw the explosion in 1961 over Siberia he was so shocked that he became convinced that nuclear arms could mean the end of humanity. He became an activist for disarmament, for peace, for human rights, for civil liberty. In his speech when he received the Nobel Prize for Peace, he called for an end of the arms race, respect for the environment, and international co-operation.

Robert Jungk was the first to get access to Los Alamos as an outsider. He wrote a book about it, *Brighter than a Thousand Suns*. He received the Alternative Nobel Prize in 1986. His concern was that the world could become an "Atomic State". His book about the Sun was published posthumously. It is a report about a new time thanks to the Sun—the Sun as the symbol of a sustainable and peaceful world.

In 1982, the Nobel Peace Prize was given to Alva Myrdal of Sweden and Alfonso Garcia Robles of Mexico. Myrdal had chaired the UN Disarmament talks from 1962 to 1973; Garcia Robles was the architect of an earlier agreement for a nuclear-free zone in Latin America. He played a key role at the 1978 UN joint programme for disarmament and was in those days lauded as Mr. Disarmament.

A very prominent member of Germany's society was C. F. von Weizsäcker. We mentioned him in the earlier chapters of this book in relation to his research on the Sun and the planets. During the war, he was also involved in nuclear research in Germany. Figure 3.1 shows the cover of his book in German published towards the end of his life: *Where Are We Going?* He addressed the questions of war and peace, of poverty and richness, of man and nature, and the value of democracy. He refers to a concept developed by his nephew Ernst Ulrich and the Lovins in the United States, "Factor 4": doubling prosperity and halving natural resource consumption. It got some importance as the motto of the so-called Wuppertal Institute in Germany.

Figure 3.1 German scientist and philosopher C. F. von Weizsäcker's book *Where Are we going?*

At this point, it is time to mention Hermann Scheer, who was since his early career a fighter against nuclear war. Later he became the father of the world's solar revolution. The following section tells why and how.

3.3 Hermann Scheer: From Disarmament to Solar Policy

Hermann Scheer (1944–2010) was a member of the Bundestag, the German federal Parliament, since 1980. He was an influential member of the SPD group there. Right from 1982, he became the SPD's speaker on disarmament and arms control. He was highly influenced in his young age by the massive demonstration for peace at Bonn in 1982 that we mentioned earlier. Years later when we had become friends, he took me to the Bonner Hofgarten and explained all what had happened at that meeting, in particular the speech of former chancellor Brandt.

In 1986, he published his first book in German, "*The Liberation from the Bomb*" (Fig. 3.2). Having a responsibility for disarmament and preservation of peace, he got convinced of the need for an alternative world, a solar world. What that would mean in practice remained first somewhat diffuse. But one thing was clear from the beginning, "no conventional energies of fossil or atomic nature, no sharing of any kind of armed conflict". Scheer was brilliant and a very kind man, but all his life he could become very nasty when it came to these important issues. The conflict with his party's coal strategy turned out to become a major problem in his career.

In 1988, Scheer created the association EUROSOLAR in Bonn. Shortly after, he came to see me in Brussels and we became very close friends. This long-lasting friendship came with a common agreement on all the strategic questions related to a comprehensive solar world, also those controversial in nature, PV and CSP, wind turbine development on-shore and off-shore, bio-energy, and storage.

Since the turn of the century, the renewable energies have made it into the mainstream of the world's energy supply and consumption. It started from Germany, which kept the leadership well into the second decade around 2013. By 2002, investments on PV had reached for the first time $1 billion in Germany; in 2017 when PV had conquered the whole world, yearly investment rose to $100 billion. **Since the turn of the century, well over $1000 billion has been invested in PV. By 2018, $3 trillion in total on all the renewables had been invested globally.**

Figure 3.2 Hermann Scheer's first book *Liberation from the Bomb*, published in 1986.

The architect of this incredible solar revolution was Hermann Scheer.

Hermann Scheer's playing field was the German Parliament, not the federal government, which never took an initiative to promote solar markets in the country, irrespective of the political constellations at any particular time. The Parliament was not a frank supporter of solar energy either. I was myself member of an Enquête Commission of the Bundestag on energy at the critical time around the turn of the century. Together with Harry Lehmann we were hardly a handful of Solar adepts in that Commission; the defenders of nuclear power were more numerous and they were a lot more aggressive. It was the art of Scheer to forge alliances across the political spectrum that made the difference. Without him, no "Solar Germany".

Scheer's first coup was the "electricity feed law", StromEinspG in German, of 1990. His party group had no majority in the

Bundestag at the time. Hence, he had to look for alliances in the conservative parties CDU/CSU. He found common interest with Manfred Lüttke and Matthias Engelsberger, who were looking for the promotion of hydropower. Daniels from the Greens was involved, too. Eventually the initiative of a law was introduced by the conservatives—normally the friends of the power utilities. And those were furious. Indeed the law that became effective on 1 January 1991 obliges them since that time to accept in the public grids the electricity from all renewables and pay for it a fixed rate, a feed-in tariff (FIT). They had to pay after this law 13.84 Pf (7 cents) per kWh of hydroelectricity and 16.61 Pf (8.5 cents) for each kWh of wind and PV electricity. The utilities went to the highest courts in Germany and the EU to complain, but without success.

The law was actually inspired by Denmark. Denmark was the first in Europe to develop a wind market. Vestas built in 1979 a first commercial 30 kW machine and later Denmark participated massively in the emerging US wind market at the Altamont Pass stimulated by an investment tax credit. On their home market, the Danes tried investment support and later, from 1988, they were the first in Europe to introduce a FIT. The Danish utilities were obliged to buy the wind electricity on offer and pay a "fair price".

Hence, Germany introduced from 1991 a FIT system like Denmark. As a result, Germany had a total of 6.1 GW of wind power installed by 2000, even more than Denmark with 2.4 GW during this period.

Unlike the success of the new wind power programme, the tariff provided in the FIT law of 1991 to PV was too low to get a market started. Until that time, PV suffered from the vicious circle: no low cost without a mass market to start an economy of scale, and no mass market without a "competitive" cost. It was Hermann Scheer's merit of breaking that circle with a new law, the "Renewable Energy Law" (EEG).

When his party, the SPD, came to power, the first thing for him to arrange was a 100,000 Roofs Programme for Germany. The draft for it had been prepared many years earlier. The Roofs programme started on 1 January 1999. It was pre-financed by the state bank KfW and approved by the finance minister of the time, O. Lafontaine. Financial support for the investments in building-integrated PV came in the form of support for the interest paid on the loan. This was by far not profitable enough for the investors. It was for this reason that the Roof programme was immediately associated with a new FIT law, the EEG. In 1999, Hermann Scheer and three colleagues of the Bundestag drafted the law. Hans-Josef Fell was one of them. They referred to the "Aachen model" of a cost-covering price paid to the owner of the PV plant. The new EEG included 45 cents/kWh in addition to the benefits drawn from the Roof Programme. The new EEG started on 1 April 2000. Immediately after came a lawsuit from the EU Commission in Brussels. It was dismissed by the European court.

The Roof programme together with the first EEG associated with it was expiring at the end of 2003. It was not renewed. A second EEG starting in August 2004 remained as the only support mechanism for PV, and wind, biomass, and hydro, which had all their special FIT tariffs. The law was unlimited in market size only since 2009. Until then the PV market volume qualifying for support was limited, 300 MW first and 750 MW later.

Hermann Scheer has written about those fascinating events at the turn of the century, of which this chapter gives only a pale report (H. Scheer. Initiating a solar revolution in Germany, in *Solar Power for the World*, W. Palz, ed., pp. 287–300, Pan Stanford Publishing, Singapore, 2014).

As a result of all this, Germany became a world leader in PV and wind power.

In 2000, Germany installed a meagre 40 MW of PV. In 2004, it had a total PV capacity of over 1 GW and surpassed Japan, which had hitherto been the traditional world leader thanks to its Sunshine programmes. Germany remained the world leader on PV in terms of yearly installation rate till 2013, when China took over. It was even surpassed in the meantime by the United States and Japan, because after Hermann Scheer's demise in 2010, the wind has turned in Germany and support is no more what it used to be.

For wind power, Germany acquired world market leadership but lost it early on again after 2005 in favour of the United States. Today China is the world leader for wind power. However, Germany still has a strong position in the world's wind industry.

Hermann Scheer got a lot of prizes and awards. In 1990, he received Germany's Order of Merit. He got the Alternative Nobel Prize and was declared Hero of the Green Century by *Time Magazine*.

Even in his early career, he was influential in his party. In his region Baden-Württemberg, he was the committee chair to decide who were the candidates to enter the voting list. Without a good place on such lists, entering the Parliament can be difficult. Since 1993, he was a member of the SPD Board and the Committees of the Bundestag on foreign affairs, agriculture, etc., and a member of the Parliamentarian Assembly of the Council of Europe.

As a student, he was the leader of the students' union in Heidelberg. This was during the student revolt of 1968 of which Heidelberg was a hotspot. Once he took me to Heidelberg to explain how he managed to keep the students at bay when they became too aggressive. "In one meeting the students voted in majority to occupy immediately the Rector's office. Hermann was against. He declared, the vote is clear, it is against the occupation. To stop any complaint, he declared, the meeting is over and got the micros and the lights on the podium cut".

3.4 The German Solar Revolution Spreading to China and the World

As soon as Germany had come forward with its political incentives for PV, China had its first major PV module manufacturer established. This happened as early as 2001. That year Shi Shenrong, a nobody in industry and finance, established the firm Suntech Power in China. By 2005, it had become the number one global producer of silicon modules. In the same year, it was registered at Wall Street in New York. Just a year later Mr. Shi had become the first dollar billionaire in PV. In those years at the beginning of the century, China had not yet a national PV programme of its own. Hence virtually all of the Suntech production was exported. I had the opportunity to visit Suntech's first production line at Wuxi, near Changhai, at the time. It was fully automated, imported from Italy.

Shi Shenrong was born in a modest family in Western China. In 1989, he went to Australia. Three years later, he got a diploma on PV from the University of New South Wales. He was a student of Prof. Martin Green, an important player in international PV development. Prof Green was well connected with PV science in the United States, Japan, and Europe; he was a faithful contributor to the big Conferences on PV I had organised throughout Europe since 1977, the EU PVSECs.

With an Australian citizenship and his master's degree in the pocket, Shi Shenrong came back to China in 2001. His company underwent like many others the big swirl of PV industry by 2012 and 2013. The company went bankrupt, but he could save a few hundred million dollars from the rest for himself, it is said.

At this point, one should also mention Frank Asbeck in Germany. His career was very similar to that of Mr. Shi. Asbeck was an absolute nobody in industry and finance, too. He had started his company SolarWorld in Bonn as soon as Hermann Scheer had created Eurosolar there. After the turn of the century, he quickly became the first German Euro billionaire from PV. His company went on the stock market and soon lost 99% of its value again. By 2017, the company had gone through two bankruptcies already bringing misery to thousands of employees who were fired. Eventually, a small version of the company was saved with the help of Qatari investors.

Suntech was not the only new PV company to emerge in the early years of this century in China—and in Taiwan. There have been a myriad of them. And virtually all of them were in the export business. We are going to see that this was going to change only around 2012–2013 when China put in place a big national market programme on its own.

By 2013, half of China's 600 big cities had at least one solar producer in place. China spent $50 billion in the decade before to build up its PV industry.

The German PV policies found many adepts throughout the world, too. By 2007, one counted 46 jurisdictions, many of them in the United States, with a FIT policy for the renewables in place. By 2012, its principle was even adopted by over 65 nations. Many studies found it to be the most effective in comparison, in particular, to the quota model and the renewable portfolio standards (RPS). The latter is an important tool of promotion in the United States; we will come back to it later.

By 2012, 13 countries had already installed more than 1 GW of PV. Seven countries had more than 4 GW installed, in this order: Germany, Italy, the United States, China, Japan, Spain, and France.

The new global PV market this century was the most spectacular. It emerged quasi out of nothing and made it within a decade or so into the mainstream of the world's power markets. But all the other renewables had their explosive growth as well. More on this in the next chapter.

Figure 3.3 Bertrand Piccard (centre), André Borschberg (right) in 2010 with the author when they won the Swiss Solar Prize. Later in 2015, the Swiss went around the world in their solar airplane Solar Impulse.

Chapter 4

Renewables Conquering the Mainstream of the World's Energy Markets

4.1 The Triumph of Solar Power

4.1.1 The World's Power Capacity from Renewable Energies Up to 2018

The most extraordinary growth of any power capacity in this new century was that of solar photovoltaics (PV). In 2018, PV has achieved a global capacity of over 450 GW; 500,000 solar panels are installed every day. In the year 2000, the capacity had achieved the first 1 GW. One GW is already a respectable amount as it stands for some 200,000 building-integrated PV installations, enough to power a big city. However, to be counted as mainstream in the world's energy markets, it should better have increased to some 100 GW. This is what happened in 2012. Since the year 2012, we have definitely seen a strong acceleration of the PV markets. In the first 12 years of the century, a total global capacity of 100 GW was reached and the 350 GW that followed needed just 6 more years to come about. Actually the yearly installation rate in these years has increased to some 70–100 GW. It is not expected that this growth rate is going to increase

The Triumph of the Sun: The Energy of the New Century
Wolfgang Palz

ISBN 978-981-4800-06-8 (Hardcover), 978-0-429-48864-1 (eBook)
www.panstanford.com

much further; we have achieved the upper limit. The recent explosion of the global markets is by and large due to the strong emerging domestic markets in China and Japan.

Until 2016, Germany had the biggest PV capacity installed worldwide thanks to its pioneering role for political incentives at the turn of the century. Currently, in 2018, Germany is fourth of the world's top PV installations after China, the United States, and Japan. China had passed the 100 GW benchmark of total PV capacity already by mid-2017.

Prices for PV on the global spot markets have fallen dramatically. Currently PV modules cost 0.30 $cents/Watt, a factor 6 or 7 from what they cost in 2010. Silicon cells that are employed on such modules come at an unbelievable low price of 0.20 $cents/Watt. This global market has been traditionally dominated by China for over a dozen years already. China and Taiwan provided in 2016 72.7% of all modules installed worldwide. The traditional PV industry in the United States, Europe, and Japan faces difficulties to compete with them. The United States has the world's second biggest PV market by now but imports most of the modules it installs; the US global market share of PV modules was a meagre 1.4% in 2016. Its main module producers First Solar and SunPower are loss making and had to lay off people. Germany does not look much better than the United States with its PV module production industry. It stood for only 2.9% of global market in 2016. Its main matador SolarWorld went bankrupt in 2018—for the third time. More on this spectacle later.

Complete PV system prices are not yet as low as they could have been, given the low module prices on the spot markets on offer. Best prices for ground-mounted turnkey installations have now come under $1 per peak Watt; for building-integrated PV, $1.3/Watt is still the minimum price to pay today.

Another big surprise for the international energy experts was wind power. True, it had already achieved 18 GW globally by the year 2000; also in the following years, it kept a bit the lead ahead of PV. Wind power had then a breathtaking acceleration and passed the global 100 GW mark already by 2008, 4 years ahead of PV. By 2011, the world had installed a total of over 200 GW. By now in 2018, this figure has almost tripled to 580

GW of global wind capacity installed. Top wind markets in 2016 were China, the United States and Germany. The world's leading wind turbine manufacturers are Vestas in Denmark, General Electric in the United States, Goldwind in China, Siemens/Gamesa, Enercon in Germany and Acciona/Nordex.

Figure 4.1 A PV array.

The United Kingdom comes as a world leader in offshore wind capacity.

Virtually out of nothing, PV and wind power, with over 1,000 GW, or 1 TW, installed in this new century, have emerged as the leaders of all the world's electricity generators.

Hydropower and biopower had a success story of their own. Renewable power electricity essentially meant hydropower in the last century. It is still a global leader in renewable power and has seen its own growth since the turn of the century. Top nations for hydro capacity are China, Brazil, and the United States.

The global leaders in biopower capacity are the United States, followed by China, Germany, India, and Japan.

The following table gives an overview on all renewable power capacities worldwide:

	Year 2000	Year 2018
Hydro	688 GW	1,150 GW
Wind	18 GW	580 GW
PV	1 GW	450 GW
Biopower	17 GW	130 GW
All RE power	733 GW	2,323 GW
All power	**3,500 GW**	**7,000 GW**

Left out here is the power capacity of a few geothermal plants in Italy and the United States and a tidal plant in France. However, at 13 GW, they are counted, too, among the renewables' total. Their role in the global picture is minor.

By now, the global capacities of coal (1,950 GW) or nuclear (391 GW) have been left behind by the renewable capacity.

The Secretary General of the World Energy Council (WEC), Dr. Frei, said about the solar triumph we see here in facts and figures: "We are beyond the tipping point of a grand energy transition".

The International Energy Agency (IEA) in Paris noticed in a report published in March 2018 (Global Energy and CO_2 Status Report 2017): "Renewables saw the highest growth of any energy source in 2017... Renewables now account for 25% of global electricity generation. ... China accounted for 40% of the combined growth in Wind and Solar PV... Nearly 40% of the increase in Hydropower was in the United States... The EU, China, and Japan accounted for 82% of global Bioenergy growth in power..."

4.1.2 A Shakespearean Drama in the PV Industry

The world's wind power industry has always been a rather quiet place. Today's top turbine manufacturers are the Danish Vestas, General Electric in the United States, Goldwind in China, the two partners Gamesa/Siemens, the German Enercon and another joint venture, Nordex/Acciona. These companies are all around since the current boom in the global wind power business started. There is not much discrepancy in the production costs;

they are well competitive with each other. And since the turn of the century, the markets have seen no such dramatic cost decrease as was the case in PV. The wind turbines became all the time bigger in size, but the price per kW installed was never too much different from what it is today. We shall see in a later chapter that dramatic technology developments did occur and a few industrial conflicts had to be overcome. But the latter were only of local and limited importance.

The PV market had a different story. First there was that tremendous euphoria in industry, in particular in Germany, that accompanied the exponential growth of the markets. By 2012, thousands of new PV companies had established themselves, manufacturers, and installers. Billions of Euros were invested in new firms, many of them the new stock market darlings. In 2010, a maximum investment of €20 billion had been reached in that single year in Germany.

All that came to an end in 2011, 2012 and 2013.

In Germany's PV industry, 110,000 jobs had been created. From 2012 to 2013, it fell by 55,000; in 2015, only 31,000 jobs had been left.

Almost all German module manufacturers went bankrupt:

- Q-Cells, the world's biggest module producer came under Korean umbrella
- Conergy
- Solon
- Centrotherm
- Sunways
- Schüko
- Scheuten, which had taken over Shell-Gelsenkirchen
- Würth, a specialist for CIS, a technology it had adopted from ZSW, the PV R&D centre in Stuttgart
- First Solar, the CdTe specialist quit Germany at that time

Bosch had bought half of Ersol in 2008 and Aleo in 2009, gave up PV in 2012 and had to write off a loss of €1 billion.

Siemens Solar gave up at the same time after a total loss on PV of €0.8 billion.

Often the stark declarations of the companies announcing that they were the new leaders of PV in the world conflicted with their poor performance in the markets—with dramatic consequences.

Some German companies survived the disaster and maintained a leading role in international PV. Wacker in Bavaria is one of the world's top silicon feedstock providers. It has kept that position since the start of the early markets until today. More on this later. SMA, the inverter specialist in Kassel, had been created by Prof. Kleinkauf, Günther Cramer and others in 1981. It has remained a global leader till today. One should also mention Solarwatt, which was saved from bankruptcy by the Quandt family. It partnered with Fronius for power conditioning, storage and other system components. Manz survived, although it lost some feathers. Phoenix Solar AG survived—until the end of 2017, when it perished for its part, too.

The PV industry in the United States, Japan and China was affected by the European crisis in 2012 and 2013 as well.

In the United States, the company Evergreen, a specialist of silicon ribbon technology went down the drain. So did ECD, the promoter of amorphous silicon on flexible substrates. The specialist of organic PV modules, Konarka took the same route out. A scandal was the insolvency of Solyndra, as it had benefited from financial support by the administration shortly before. A latecomer was Sun Edison, which perished in 2016.

In Japan, a traditional leader in the global PV industry, companies such as Kyocera, Sharp, Showa Shell, Panasonic and Solar Frontier felt also a tsunami from that European earthquake but were never affected in their existence. Japan, which was an exporting nation for PV modules until 2012, became a net importer, at the expense of its own industry.

In China, the global pioneer Suntech Power, which we reported about earlier, perished, too, in 2012.

Interesting is the story of a few PV industry old-timers that disappeared from the global PV radars in that terrible period. Germany had Schott Solar with a lot of different roots. Until it gave definitely up in 2012 and 2013, W. Hoffmann, also the long-time chairman of the European PV Industry Association (EPIA), managed its fate towards the end. The origins go back until 1958 to the famous AEG-Telefunken, later Deutsche Aerospace, DASA. In 1979 was created RWE-Nukem, a thin-film module producer. In the same year, MBB, the aerospace company in Munich, and Total Energy set up Phototronics specialising in amorphous silicon modules. In 1994, all the three, DASA, Phototronics, and Nukem became a joint company, the new ASE under the roof of RWE, Germany's nuclear energy champion. In 2002, Schott entered the scene. Schott is a German glass manufacturer. Three years later, Schott RWE Solar became a wholly owned subsidiary of Schott only. In 2012, it gave up its silicon activities and a year later those in thin films.

In the United States, the early PV business was actually centred on the two oil majors: BP and Shell.

On the one hand, there was the subsidiary of an oil company, ARCO Solar, existing since 1977, which was sold to Siemens in 1989. In 2001, Siemens Solar was sold to Shell. Shell gave up its PV business in 2009. Only in 2016, it made a statement it might come back with a New Energy Division.

On the other hand, there was BP Solar. This goes actually back to Solarex, the new PV leader established in 1973 by my friends Joseph Lindmayer and Peter Varadi. Solarex had absorbed an earlier start-up, the Solar Power Corp., created by Berman in 1969 and sold its shares in 1983 to another oil company, Amoco. Not because it was insolvent, but because the owners wanted to reap some benefits. Later, Amoco merged with BP Solar. And as mentioned earlier, BP gave up PV in 2011—only to manifest new interest again in 2017.

A special case in all that chaos is SolarWorld. It was established by Frank Asbeck in Bonn in 1988. And it went down the drain like all the other traditional module manufacturers by 2013. It missed bankruptcy only through the intervention of Qatari investors as mentioned earlier. The company had lost 99% of its stock market value: It went from €5.3 billion in 2007 eventually to 13 million in 2017. The difference in this case is that SolarWorld initiated its own PV association, ProSun. What followed was a political initiative against the Chinese competitors in the field of silicon modules. As SolarWorld has next to its plants in Germany strong operations in the United States, it mobilised, via its association ProSun, both the authorities in Washington and in Brussels against the Chinese module producers, saying that competition was unfair. Many in the field think that this claim is unfair. Whatever, SolarWorld was successful and punitive tariffs are now applied on the import of Chinese modules into the United States and Europe.

Eventually these punitive tariffs were confirmed. By mid-2017, the International Trade Commission (ITC) found injury to the domestic crystalline silicon cell industry based on a petition brought by Sinova and SolarWorld. Thereupon, the president of the US Solar Energy Industries Association (SEIA) declared on 22 September 2017: "The ITC decision is disappointing for nearly 9,000 US solar companies and the 260,000 Americans they employ". And "Foreign-owned companies that brought business failures on themselves are attempting to exploit American trade laws to gain a bailout for their bad investments". And "While we believe that this is a bad decision based on Suniva and SolarWorld mismanagement, we respect the commission's vote".

It did not help SolarWorld and its associates at all as it went bankrupt again in 2017. Its products are just not competitive in quality and price. But it hurt the markets here. Module prices are artificially increased by the duties, and to some degree this is a discouragement for potential investors. Politicians must understand that the modules that are in use are the imported ones, no matter the duties. In early 2018, the US president applied with much publicity new import tariffs on solar modules. They were also applied to SolarWorld modules. Finally, in March 2018, the company went broke a third time. Will it be the last time?

The 10 biggest module producers on the global markets are currently Trina Solar, Jinko Solar, Canadian Solar, JA Solar, Hanwa Q-Cells, CGLSI, First Solar, Yingly Green Energy, LONGI Green Energy Technologies, and EGing Photovoltaics. Yearly production capacity of the six largest of them is around 6 GW each. Except one, all of them employ crystalline silicon square cells. All companies are Chinese except three. It is not unusual that companies have their manufacturing plants also outside China, or the three other home countries, the United States for First Solar, Canada for Canadian Solar, or South Korea for Q-Cells. Productions are indeed run to some extent in Malaysia, Vietnam, Thailand, South Africa, Brazil, Portugal, and the Netherlands.

The stock market value of the most valuable PV companies is given on the PPVX of Top 50 Solar. Since 2003, the total value has increased by 269%. But between then and now, there have been impressive ups and downs. By 2017, the Sun was shining again on that PV stock market. The total value of the 30 listed international companies was €23.9 billion. It grew for the first 8 months of 2017 by 7.9%. The top runners were the Israeli SolarEdge Technologies (+115%), the Swiss Meyer Burger Technologies (+115%), the Chinese Jinko Solar Holding (+73%), the US First Solar (+49%), the other US SunPower (+40%), the German SMA (+38%), and a few others.

In conclusion, the disaster that happened around 2012 crushed tens of billions invested in the PV business in Germany and the associated markets. Over 100,000 jobs were lost in Europe. Since those events, PV lost some attraction in Germany and Europe. Many investors got their fingers burnt and confidence in PV has not yet totally recovered.

The situation is really absurd. Germany, Spain, and Italy were the pioneers of PV's global market development—at the expense of the electricity consumers in these countries. At the beginning, there was a hefty bill that had to be paid. At the start of an economy of scale, costs are high. In the first decade of the century, the European consumers spent the necessary hundreds of billions of Euros. And now that costs have come down, the Europeans have sidestepped the market. Germany installed less than 2 GW anew in 2016, Europe hardly 7 GW, and China installed 24 GW in 6 months.

What had been the crucial trigger of these events in 2011, 2012, and 2013? As mentioned before, the German EEG in its original version of 2004 had imposed a market volume for those benefiting from the generous FIT. Only by 2009 the FIT market volumes in Germany became unlimited. Consequently, in 2009 Germany installed 4.5 GW of new PV capacity, and in the following 3 years, some 7.5 GW were newly installed each year. At 26 TWh fed into the grids by 2012, the yearly disbursement to the investors on behalf of the FIT was dangerously approaching the €10 billion mark. Berlin reacted in 2012. The FIT was reduced by 40% between 2012 and 2013. In retrospect, one must say this was ok. What was not ok was that a new upper volume of PV installations per year was imposed. And that brought the market almost to a halt, and the industry into the abyss.

Spain, the other frontrunner, reacted similarly. It stopped its generous FIT system completely. Since 2012, no new PV was installed above the 4.5 GW it had already. Only in 2017, some new PV GW came on the table by a new auction system.

4.2 Renewable Energies for Heating and Transport

4.2.1 Bio-Energy, the All-Rounder

Unlike photovoltaics, where one single component, the solar cell, is replicated a 100 billion times and spread all around the world, bio-energy is something a lot more complex. Biomass from which it is derived comes in the form of products of forestry, agriculture, cattle breeding, and waste streams originating from the production and consumption of our food. Conversion of the different forms of biomass is complex, too. There are various ways of transforming the biomass into an energy product, biological or thermal. Transformation of biomass can create pollution. In some cases special precautions are necessary to prevent it. For commercial systems, Europe imposes since 2018 strict exhaust gas limits. Also to eliminate dust, cyclones and various filters become mandatory.

Biomass generation concerns the whole biosphere. The green matter around us absorbs the Sun's incident radiation. The matter and energy created in this way end all up in thermal decomposition. The generation and use of bio-energy is the art to catch some of the energy before that happens.

Bio-energy is a vast field. We published a special 700-page book on it in 2015: *Biomass Power for the World: Transformations to Effective Use* (edited by W. van Swaaij, S. Kersten, and W. Palz, published by Pan Stanford Publishing, Singapore).

Fourteen percent of the world's energy consumption today is met by bio-energy. It is the biggest among the renewable energies. A major part is consumed as fuel wood or charcoal for cooking purposes in less developed countries.

Global bio-energy production and utilisation involved in 2015 a budget of $674 billion; that was 15% more than in 2010, 5 years earlier.

Figure 4.2 Our book on bio-energy from 2015.

Liquid biofuels, biogas, and wood pellets are like PV and wind power newcomers in our century, leaving alone that China has a long tradition for biogas and Brazil for the production and use of bio-ethanol for transport. Unlike PV and wind power, bio-energy comes in its various forms as a stored energy and that makes it particularly suitable for combination with them.

The following table gives an overview of the main forms of modern bio-energy by 2018 (in litres and million tonnes of oil equivalent [Mtoe]).

	Market volume	Market value
Biofuel liquids	155 billion litres or 115 Mtoe	$103 billion
–Ethanol	–120 billion litres	–$72 billion
–Biodiesel	–35 billion litres	–$31 billion
Biogas	20 Mtoe	$15 billion
Wood pellets	12 Mtoe	$6 billion

These figures are by and large valid for the industrialised countries. If one includes China, too, they come considerably higher in the case of biogas. With its tens of millions of digesters in place, China's biogas market might be more than a hundred times larger than that in the rest of the world.

The economic importance of these modern forms of bio-energy is impressive. Although, as mentioned earlier, they started their explosive growth only with the beginning of our new century, they are part of our move into the solar century that we enjoy by now.

Examples are the **biofuels for transport** that started from a meagre 10 Mtoe in 2000: By now they have passed well the 100 Mtoe mark and represent with over $100 billion per year an important economic factor.

Some 60% of the world's **alcohol** for transport is produced and consumed in the United States. Only 1% was the share of ethanol in US gasoline by the turn of the century; now the standard is E10 with 10%. It could be more, but there is what is called a blend wall: Car engines would have to be modified to tolerate a higher degree. And there is a lack of infrastructure for supplying E15, for instance. E85 with 85% of bio-alcohol is used in the US Midwest, where the big production plants for ethanol are located. Over 10 million vehicles that can run on

E85 are in use there. The US market for ethanol is currently not growing too much anymore. However, there is still in the background the mandate of the Renewable Fuel Standard from 2007. It projects an increase of consumption from 15 billion litres in 2006 to 136 billion litres or 36 billion gallons in 2022.

Brazil produces one-fourth of the world's ethanol. It is traditionally the country with the highest rate of national utilisation. It goes back to national programmes in the 1970s that were triggered by the global oil price crisis of 1973. In those years, I personally had an opportunity to follow its birth more closely when I was a member of a French governmental mission that visited the country to discuss possible co-operation.

Nowadays, some 50% of the fuel for light-vehicle transport in Brazil stems from nationally produced bio-alcohol. It is not subsidised, leave alone that it is not as highly taxed as the conventional fuels. It is a major part of the national industry of sugar cane in the south of the country. Fifty-three percent of the income of that one is generated from sugar and 42% from the ethanol it produces. Bio-ethanol is very job intensive: In Brazil 3 million people are employed in the sector.

Bio-alcohol for transport is used in two forms: E27 and E100. E27 is a mandatory blend imposed by the national government, with 27% of anhydrous ethanol in regular gasoline. E100 that is called the gasohol consists of hydrous alcohol. Associated with the alcohol programme, Brazil saw an industrial revolution in the auto market. The year 2003 saw the introduction of flexible fuel vehicles (FFVs). Today Brazil has 25 million of them and virtually no other types are left on the roads. Conventional car importers such as VW and Ford had to adapt. FFVs can run on any ethanol blend, be it E27 or E100. People buy at the filling station what is locally on offer as the cheapest.

The EU has a mandatory target since 2009. It is part of the directive 2009/28/EC. The Brussels directive is binding for all 28 EU member countries. As it stands, the target of 10% biofuels for transport by 2020 is not going to be reached. A particularity of the European car market is the predominance of diesel over gasoline—which is part of the "diesel scandal" story of the European car manufacturers since 2016. Anyway, Europe employs at 6.6% almost twice as much biodiesel as it does for ethanol with 3.4%.

Bio-ethanol is now employed worldwide in transport. A mandatory share is imposed in more than 30 other countries also. Even if it does not come specifically at the filling station as E10, E85 or E100, virtually all gasoline has a small percentage of alcohol blended with it, be it only to increase the octane number.

Biogas is another important sector of bio-energy. China is a world leader with 80,000 agricultural and millions of rural family digesters in place. Most of them were installed after the turn of the century only. Other big producers are Germany, the United States and the United Kingdom. Europe operates 17,500 biogas plants. In comparison, the United States has installed 2,200 plants mostly for waste treatment.

Germany plays a particular role from several points of view. In Europe, it is by far the largest producer with 11,000 plants in operation by 2018. Germany's development of biogas started with the EEG initiatives mentioned in connection with PV and wind power. Moreover, Germany is the particular promoter of the digestion of cereals with the traditional manure and other agricultural wastes. The innovation was to feed the whole plants like maize into the digester from the root up to the fruit. In Germany, plants for digestion purposes are grown on 1,400 ha of agricultural land.

Germany has 44,000 jobs in the bio-energy sector. It has a power capacity of 4.2 GW with biogas plants, most of them CHP (combined heat and power) plants. Biogas creates a yearly value of €8.3 billion. Some plants have the facility to extract methane from the biogas for injection into the gas networks. Biogas is also suitable for use in transport.

The FIT that is paid in the frame of the EEG comes at 16.9 cents/kWh. This generous tariff is explained by the particular attraction of biogas for providing a continuous supply in combination with the intermittent availability of solar or wind. It comes also as a support for Germany's agriculture. Besides, agriculture also profits from the PV installations deployed on large farmhouses' roof areas.

In January 2017, a new regulation entered into force, the new EEG 2017. For biogas, like for the other renewables, Berlin introduced an auction system. However, the profitability of biogas production remains untouched. What is new is the corridor that limits the number of new installations to some 200 per year. Also limited is the addition of cereals to 50% of all feedstock.

The global **wood pellet** market is rather recent, too. The production capacity stands at 42 million tonnes. Mills have been counted in 21 countries, but most of the world's pellets come from North America. Pellets are employed in the industrial sector for heating instead of heating oil, for district heating. The power sector is the fastest growing. **Co-firing** in coal boilers is the preferred new market. In the United Kingdom, the big coal power station Drax got two-thirds of its capacity converted to pellets.

The cost of electricity from co-firing is attractive. It comes relatively low at 2 cents/kWh when local feedstock is available. For industrial purposes, pellet prices achieve some $150 per tonne.

Europe employs 20 million tonnes of pellets a year. It is the world's largest consumption market; a third of it is imported, mostly from North America.

4.2.2 Solar Heat Collectors

It's again the same story: Like the other renewables, solar heating developed virtually out of nothing in this new century. By the turn of the century, the world had installed a total of just 70 km^2. In 2018, the total thermal capacity has reached a collection area of some 820 km^2. Specific yields in Europe are 575 kWh of heat at 50°C and 411 kWh at 75°C per m^2 and year for a flat-plate collector. For tubular, collectors it is more.

These 820 km^2 of solar heat collectors compare with an overall area of some 3,000 km^2 deployed to generate the 450 GW of PV peak power we have now in operation, as mentioned earlier.

Over 120 million solar heating systems are in operation worldwide. Seventy percent of the world market is in China, followed by the EU and North America. China has the products of the highest value, the vacuum tube collectors on offer and in operation. China's systems are also the most competitive compared to the flat-plate collectors that dominate the rest of the global markets. Ninety-four percent of the collectors are employed in solar domestic hot water systems for single-family houses.

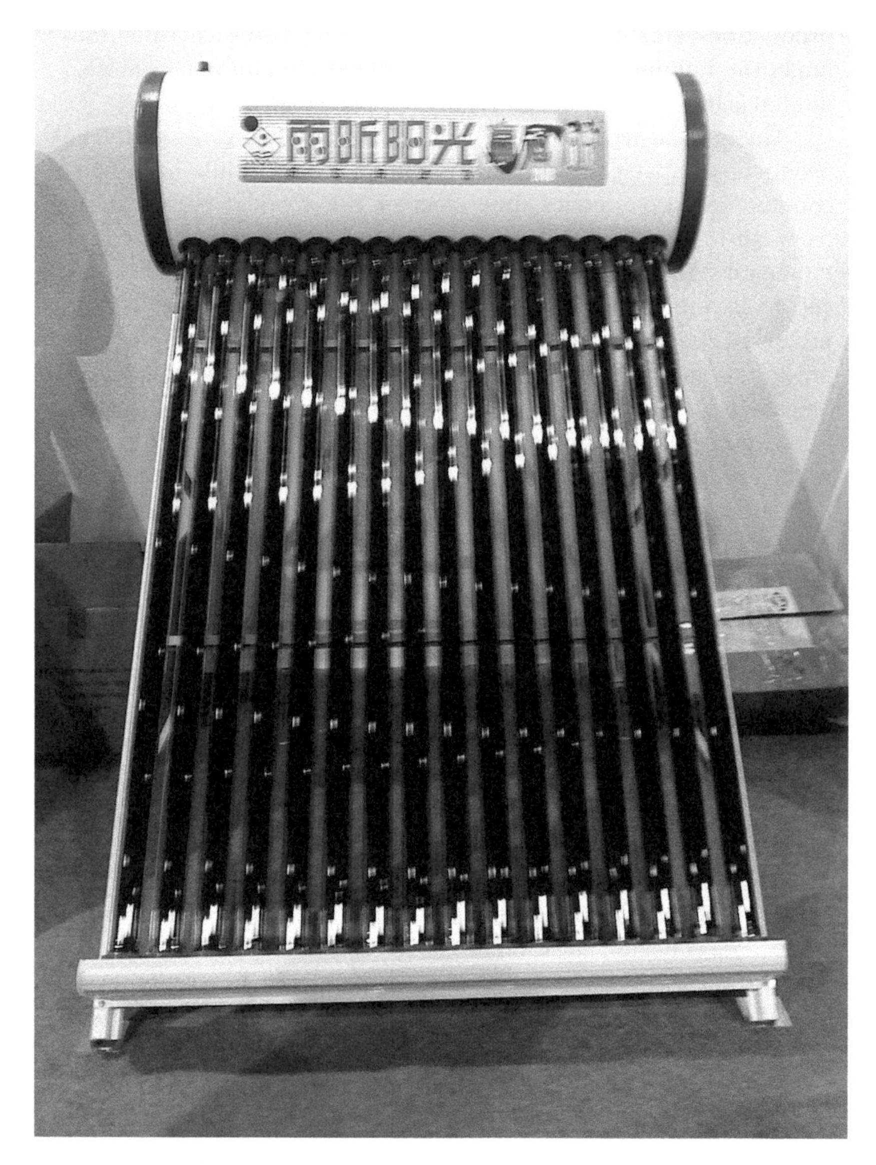

Figure 4.3 A Chinese solar water heater (picture by the author).

Denmark has installed the world's largest district heating system end of 2016. It has a solar collector area of 157,000 m^2 or 15.7 ha.

The yearly turnover of the global solar heating sector is a respectable \$25 billion. It involves 730,000 jobs.

4.3 A Summary of Global Achievements

From what we have analysed so far, we can draw the following conclusions:

- The renewable power capacities led by hydropower, wind power and PV have by now conquered the mainstream of the world's power markets. They passed the global capacities installed from fossil sources, led by coal and nuclear.
- There is no ideology involved: The conventional powers lost the favours of the investors as the solar newcomers were beating them on their own field, market cost.
- This growth started only some 18 years ago at the turn of the century. This industrial revolution involved thousands of billions of US dollars. It has turned our young century into a "Solar Century". Most people have not yet realised: We have not only entered, we are already right in the middle of a "Solar Age".
- Besides the power sector, modern bio-energy and solar thermal saw their markets expanding with a similar explosive growth.
- Thanks to the renewable energies, there is higher comfort in the building sector while energy consumption is reduced. This trend is going to be reinforced in the future. An EU building Directive imposes the obligation to make all new buildings in Europe from 2021 "Near-Zero-Energy-Buildings". That means to apply systematically PV, storage, and other renewables in the building and on its envelope. California tried in 2017 to make PV mandatory on all new buildings, but the draft bill could not get approval. However, eventually things changed again: On 9 May 2018, California decided to make PV on all new houses mandatory from 2020 onwards.
- The renewables' development was initiated by particular policies in Germany. In the meantime, market leadership has moved to China followed by the United States.
- Energy consumption comes down. In Europe, it shrank by 7% since 2000. CO_2 emissions are also getting under control. They stabilised globally since 2013 despite an economic growth of 3%.
- Renewable energy policies were complemented by specific support schemes such as the FITs, or the US PTC, the

Production Tax Credits, applicable to wind power, or the ITC, the Investment Tax Credits in support of PV, or the RPS, the Renewable Portfolio Standards of the US States, the Green Bonds, and a lot of others.

- The initial financial support for stimulating the new energies' emerging markets kept always below the big subsidies in place for the fossil and nuclear energies. Besides the hundreds of billion dollars given every year in support by the G20 nations, some US States gave and give subsidies to their coal industry even before the new US president arrived: In 2013 they gave $20 billion for coal; other States give out-of-market support to avoid collapse of their nuclear power plants' operation.
- Since the turn of the century, 10 million new jobs in the renewables sector have been created virtually out of nothing. By 2030, they could reach 24 million, one-third of them in China. In the United States, one estimates that one direct solar job induces two additional jobs elsewhere. The 260,000 direct jobs in solar PV correspond actually to some 790,000 in total.

The corresponding economic output is estimated at $154 billion a year.

And a last word about finance and competition.

By 2017, new coal power plants generated electricity at some 7 to 10 cents/kWh compared to wind electricity at 5 cents/kWh or less and solar electricity from PV plants at 7 to 8 cents/kWh in Central Europe and less elsewhere.

The total park of 580 GW of wind power installed globally produces some 1,160 TWh of electricity this year. This electricity generates at 5 cents/kWh a value of **$58 billion**. On the other hand, for extending the park, counting $1,600/kW per wind turbine installed, the total disbursement for the global installations of 50 GW by 2018 then comes out to be **$80 billion**.

The 450 GW of PV power installed globally by 2018 produces a value of **$46 billion** under the assumption of a production of 540 TWh of electricity and a kWh price of 8 cents. Always with the idea to extend the park with new PV power, this can be compared with an investment of **$100 billion** for 75 GW at an average investment price of $1,300/kW that year.

Otherwise expressed, in a few years when, on the one hand, the world's parks of solar PV and wind power will have grown further and, on the other, the installation costs of the systems will have come down even more, the productive investments of the new solar and wind power plants will become equivalent to the income from the production of the existing parks. Financially speaking, the running parks generate the money to install more and more of the plants. They become their own breeders. The idea works as long as the lifetime exceeds well the 10 years that are the basis for the kWh calculations. However, there is no doubt about that PV generators have a life of well over 40 years and for wind turbines one counts currently more than 30 years.

4.4 Renewables around the World

4.4.1 China

China's power capacity of 1,649 GW in 2016 was the highest worldwide. It also has the highest capacity of coal power.

As mentioned earlier, China is the world champion in the renewables' electricity markets for hydropower, wind power and PV.

China's hydro capacity stands for 20% of its total capacity. It has doubled since 2007.

In wind power, it went from 1 GW in 2000 to 188 GW by 2017. For support, a national FIT system has been applied since 2009. It was revised lately and the following tariffs are valid since 2017: Depending on the four regions into which China is divided for the FITs, the new tariffs range from 6 cents to 8.5 cents per kWh. For the first time, a special tariff of 12.7 cents/kWh for offshore wind electricity was introduced.

China's PV market grew recently at an extraordinary speed. It passed cumulated PV installations of Germany only in 2016. Then its PV capacity also passed its own nuclear capacity it has in operation. A year later in 2017, China's PV capacity had already passed the 130 GW mark and tripled Germany's total.

Figure 4.4 A book about China's wind power from 2010, *Energy Revolution: Possibility of Wind Power*, by the Chinese Wind Power Association.

Since 2008, China is globally the greatest PV module producer. However, in the early years, almost all was exported as its domestic market passed the 1 GW mark only in 2011 for the first time—more than 10 years after Germany did. And yet colourful PV programmes were not missing. China had in the 1990s a "Brightness" and a "Township electrification programme", but markets remained minuscule. Since 2006, China had a Renewable Energy Law in place fixing targets and mandatory grid connection and purchase. In 2010, the FIT was raised to RMB 1 (15 $cents) per kWh backed by a nationwide enthusiasm for solar PV. This made investments profitable and the markets started off. Like for wind power, the FITs were revised in 2017. The feared strong reduction did not materialise, an encouragement for the continuation of the PV market to blossom. The latest FITs range from 9.4 to 12.9 cents/kWh.

Figure 4.5 Beijing, 2015. Public presentation of the author's book *Power for the World* in Chinese. From the left: Gao Jifan, president of Trina and president of China's PV Industry Association, the author, and his friend Qin Haiyan, president of China Certification for RE. At the microphone, Li Junfeng, director of China RE Association and DG of National Centre of Climate Change Strategy.

China is a vast country and full electrification was not achieved until 2015. From 2013 to 2015, $4 billion was invested to supply electricity to the last 2.73 million people who were still lacking it, in particular in Tibet. The cheapest on offer was 0.5 to 1 kW of PV with an investment of $1,400 to $3,200 per household.

As mentioned earlier, China is the world's leader for biogas generation and use and for solar heating, too.

By the end of 2017, China had an overall capacity of over 500 million m^2 of solar heat collectors installed. This represents some 80% of the world's total. In 2016, it had installed 34 million m^2 of new vacuum tube collectors and 5.4 million m^2 of new flat-plate heat collectors. Some 70% of the collectors are employed for solar hot water (SHW) systems, the rest for space heating and cooling and applications in industry and agriculture. A SHW with tubular collectors costs in China approximately $200 per m^2 of collector area.

Figure 4.6 Beijing, 2015. The author with vice minister of the Ministry of Industry and Information Technology (MIIT), Government of China, the Honourable Dr. Huai Jinpeng.

For biogas in China, an overview was given by Li Xiujin from the Beijing University of Chemical Technology. After Li, biogas was promoted by the government since the 1970s. In the 10 years between 2003 and 2013, 42 million household digesters were installed, each with a capacity of 8 to 12 m^3. One hundred sixty million people in China's rural areas benefit from them. Besides that, 80,000 units were installed by animal farmers. Some 15 billion m^3 of biogas is produced every year. In total, the national government gave a support of $5 billion. The sector employs 290,000 people and counts over a thousand production and service companies.

4.4.2 Renewable Energies in the United States

In 2018, the power capacity from renewable sources has a share of 20% of the total US capacity. In the past decade, the part of hydro-, wind-, and solar PV power increased all the time, while that of the conventional capacities from coal, nuclear and oil declined. In 2016, $46 billion of investments into 21.5 GW of new electric capacities of PV and wind power, 90% of all investments in the sector, concerned those of renewable energies. Further, 2017 was the second year in a row when PV and wind power investments made up more than 60% of new installations, beating natural gas, which came second. For coal plants, there were only retirements instead of growth. Nor does it look good for nuclear. We have already reported about the disaster there in a previous section. A better example is the utility Duke Energy, which cancelled a nuclear project in Florida in 2017 and announced to invest $6 billion instead in clean energies, 700 MW of PV and some storage capacity.

The account is equally impressive in terms of job creation; in the PV and wind power sectors, 475,000 new direct employments have been created in the last few years leaving 187,000 for the conventional power sector.

Hand in hand with the emergence of the renewables went the reversal of the emission growth of GHG; in 2016, GHG emissions were the lowest in 25 years.

The triumphal march of the renewables started with wind power. The United States was the first nation to systematically promote its deployment, from the 1980s onward. Well known are the first commercial wind parks in California at the Altamont pass and a few similar sites. At the end of 2001, 4.1 GW of wind power had been installed in total, but that was only the beginning. Since then, 52,000 turbines have been installed. Total wind power capacity passed 90 GW in 2018. Wind turbines on the US market benefited from a 66% price reduction since 2009. Some $250 million of lease payments go to landowners, farmers and ranchers. In 2016, $16 billion was invested in new wind parks. Wind provides now over 5.5% of US electricity generation. The kWh price comes today at 5 cents. Leading US states for wind power are nowadays Texas, Iowa, Oklahoma and California. The estimates by the National Renewable Energy Laboratory (NREL) come forward with possibly 200 GW of wind power installed in the United States by 2030 delivering electricity at 2.3 cents/kWh. Not sure that offshore, which was a subject of heated debates for many years, can play a role then. By 2016, only 60 MW of offshore was in operation.

Next to wind power, the United States has a thriving solar PV market. Actually it started slowly and passed Germany's installed total only in 2016. In 2000, the United States had only 4 MW of PV installed, and in 2010, the capacity increased to 851 MW, but the 1 GW mark was only passed in 2011. In 2017, its total capacity exceeded 50 GW, the world's second largest after China. In 2016, 39% of new capacity installed in the country was PV power, more than new natural gas power and wind power. Leading US States for PV are California, North Carolina, Arizona, Nevada, New Jersey, Utah, and Hawaii.

The United States has 1.3 million building-integrated PV installations in place. Lawrence Berkeley National Lab runs a national data set on PV. Approximately 50% of the PV systems are customer owned. There are large solar leasing companies. Energy Sage is one of the important online market places for solar, leveraged by National Grid, Sierra Club, WWF, and Staples, with information on over 350 solar installation companies. Connecticut Green Bank launched GoSolarCT.com. There are a few other ways of promotion. The Time of Use (TOU) rate was introduced to widen the availability of solar electricity during the day with storage. There is community solar, rooftop leasing and expanding access to low-income customers.

Google has analysed 60 million buildings in the United States. It concluded in a publication in March 2017 that nearly 80% of US rooftops are suitable for PV array installation. Houston has the highest solar potential on its buildings with a potential of 18.9 GWh production per year, ahead of LA, Phoenix, San Antonio and NY City. With data from Google Maps and Google Earth, it did 3D modelling and developed a "tool search function" to find the potential of individual roofs for PV. They also got the "Project Sunroof", which advises on the best array size and estimates the energy produced and how much it costs to lease the panels.

Another trend goes towards PV utility scale; in 2016, 80% of all new capacity was indeed of big utility scale. In 2017, prices for utility-scale PV plants came for the first time under $1,000 per kW installed, turnkey. Some utilities conclude PPA (more on the PPA in a moment) 5.5 cents/kWh to 5 cents/kWh. The lowest in 2017 was one with Tucson Electric Power for less than 3 cents/kWh of PV electricity.

America was the first to put in place a groundbreaking tool to get renewable electricity markets off the ground. In 1978, long before Denmark, Germany, or China did, was decided the federal regulation that utilities became obliged to accept on their grids the electricity offered by RE generators and to pay a "fair price" for it. The Public Utilities Regulatory Policies Act (PURPA) was introduced that year and by 1991 opened to all sizes of generation.

However, the markets did not move. What went wrong? The explosive growth of the global markets started only later when Germany and the many countries that followed had put in place the EEG and the FIT. Those made the difference. However, the "fair price" followed the point of interest of the grid operators that do not pay more for the RE electricity than they would pay for the regular conventional one. The EEG follows the point of interest of the producers that want a cost-covering price. In the early days, the cost of PV, wind electricity, etc., was still very high and so was the difference between the "fair price" offered by the utilities and the cost-covering one demanded by the investors and paid eventually by the EEG or the FIT.

For financial promotion, the United States does not use the FIT, which is the popular instrument in many other countries. The principle of the FIT is that all customers of the electricity supplier together pay the tariff that is disbursed in support of the investors in PV, wind power, etc. The United States employs rather the tax credits. That means that not the electric utility customers but the taxpayers pay for the investments. The corresponding federal PTC, a corporate Production Tax Credit, was first enacted in 1992. It stood at 1.5 cents/kWh in 1993 and was renewed and expanded numerous times. An equivalent cash grant was possible. The PTC is still in place today. The rate is just augmented to 2.3 cents/kWh to take into account the inflation rate since the 1990s. This support is not very generous, more so as the rates would be further reduced in the coming years and the support would be stopped after 10 years. In practice, it is mostly applied for wind electricity. We notice the same phenomenon that promoted wind power in Germany since 1991 with a new feed-in law: It was a welcome support for wind power but not profitable enough for PV.

Since 2015, the US PV markets in the residential and commercial sectors benefit from a 30% ITC. The applicability of the ITC will range into 2021. Its origin goes back to 2005, an "Energy Policy Act" of the time.

The American Council on RE (ACORE), noticed recently that the renewables in the United States were the largest source of private sector infrastructure investment over the past 6 years; $100 billion was raised alone in 2015 and 2016 by RE tax credits. However, it also noticed that bio-energy, hydropower, geothermal power or fuel cells were excluded from the 2015 tax agreements.

In a report of mid-July 2017, the Lawrence Berkeley Lab published a report on the RPS. The US States have a support mechanism of their own, the RPS. It is noticed that roughly half of all growth of renewable electricity in the United States since 2000 is associated with the RPS. Nationally, RPS represented 44% of all new RE power additions in 2016. The cost of REC, the RE Certificates issued to meet the general RPS obligations fell in 2016, another encouragement. Interim RPS targets are currently met. Targets were even increased in District of Columbia, Maryland, Michigan, New York, Rhode Island and Oregon.

The power purchase agreements (PPA) are a means to purchase RE directly from the generators at a fixed price. In the US market of 2016, half of the newly installed power capacity involved

PPA. Companies such as Google and Amazon use them to improve their green credentials.

In this context come also the yield cos. In short, the purpose is to invest in companies, the yield cos, that own a portfolio of operating RE projects. The income is stable and foreseeable. That makes a difference with manufacturers and installers. We saw earlier that the stock market values of RE manufacturing and installation companies can be quite erratic. The yield cos provide a low-risk opportunity for investors. They are of interest for institutional investors and individuals alike.

There is also a newcomer, the green bonds. They came into use to raise funds for clean energy projects. In 2007, they did not even exist, but by 2016 green bonds worth $182 billion had been issued by communities, enterprises such as Apple and Toyota, banks and states. Recently, France issued its first green bonds. At 17% of the global total, the United States comes second after China, followed by Holland, Germany and India.

In the following, we come back briefly to **California**, which has a special dynamism for RE in the United States. It wants to become a RE super power. In a Clean Energy Act of 2017, California declared its determination to become 100% RE by 2045. By 2020, a third of its electricity is planned to come from the renewable sources. The RPS benchmark stands at 60% RE for 2030. California's carbon intensity has reduced by 40% since 1990.

Currently, half of new US PV power is installed in California. Many cities have regulations for mandatory installation of PV on new buildings. Going with this trend, on 9 May 2018, California decided to make PV on all new houses mandatory from 2020 onwards. There is three times more PV capacity in operation than wind power. Total PV power passed 25 GW in 2017. Until 2003, it was zero.

Over $50 billion has been invested so far in PV in California. It involves 100,000 jobs, actually 236,000 jobs including the indirect ones; $16 billion a year is paid out in salaries, wages and benefits. California has 500 PV manufacturers and over a thousand installers.

An equivalent of 4.7 million homes are supplied with PV electricity. California's biggest PV plant is the 579 MW Solar Star, built by SunPower near Rosamond.

Before concluding on the situation in the United States, one should not forget to mention its global leadership on bio-alcohol for transport. Details are given in a preceding chapter.

In 2016, we published a special book on RE in the United States: *The US Government & Renewable Energy: A Winding Road,* by Allan Hoffman, also with Pan Stanford Publishing, Singapore.

4.4.3 Germany

In 2017, over 15% of Germany's final energy consumption was met with the renewables, three times more than at the turn of the century. Two-thirds of these renewables were derived from biomass. This high percentage of the biomass is in line with the situation in the rest of the EU and the other industrialised countries as well.

Among the German renewables, the share of the electricity generation benefited from an extraordinary explosive growth; but their share in the heating sector only doubled since the year 2000 and in the transport sector renewables' part did not move very much at all.

For 2017, final figures for the German **electricity market** are available. By then, 202 GW of total capacity was in operation, producing 550 TWh of electricity. That makes on average only 2,720 hours of equivalent operation (called the capacity factor) for the full year. In terms of **capacity**, all renewables together stood at **55% of the total**:

• Wind	55.3 GW
• PV	42.7 GW
• Biopower	7.4 GW
• Hydro	5.6 GW

The conventional power capacities were the following:

• Natural gas power plants	29.5 GW
• Hard coal	25 GW
• Lignite	21.3 GW
• Nuclear	10.8 GW

For effective electricity **generation**, the conventionals were in 2017 at the top with 62% of the total against **38% for all renewables** in the following order (the equivalent operational time in 2017 is shown within parentheses):

• Lignite	136 TWh (6,380 hours)
• Wind	101 TWh (1,830 hours)
• Hard coal	84 TWh (3,360 hours)
• Nuclear	72 TWh (6,670 hours)
• Bio-electricity	48 TWh (6,490 hours)
• Natural gas power plants	46 TWh (1,560 hours)
• PV	38 TWh (890 hours)
• Hydro	21 TWh (3,750 hours)

In summary, the capacities of wind power and that of solar PV beat all the conventional ones—a remarkable fact as in 2000 yet they counted for almost nothing! Even in terms of production, wind electricity in 2017 for the first time beat even the electricity generated from hard coal and that from nuclear. Hard coal consumption was losing ground; its consumption in power plants decreased by 16% compared to the preceding years (figures from FhG ISE). Eventually, Germany's CO_2 balance sheet is improving accordingly.

Electricity from renewable sources as part of overall electricity consumption grew from just 6% in 2000 to 38% in 2017. As we saw, almost half of the RE electricity comes now from wind power. The second block is bioelectricity with hydro. The latter complement very nicely the intermittent powers of wind and PV to guarantee continuous supply. The biopower capacity is much smaller than that of wind, but that is compensated in part by a longer operation time in the year in the case of biopower.

PV power saw a spectacular market success: Until 2016, Germany had the world's largest PV power park before being passed by China, the United States, and Japan. In 2017, PV's share of electricity supply in Germany was 7.2%.

Germans like their renewables: 95% of the population is in support. All would agree to have PV in their neighbourhood, even for wind turbines the "nimby—not in my backyard" proportion is under 50%.

Germany has 330,000 jobs in RE.

In 2016, €15.2 billion were invested in Germany in the RE sector, most of it in new wind power.

The renewables benefit from financial support. It was calculated that Germany's amounted to €54 billion between

1970 and 2012. However, it was actually minuscule compared to the subsidies given in the same period of time to atomic power (€187 billion), power from hard coal (€177 billion), and lignite (€65 billion). In 2017, state subsidies out of the pocket of taxpayers for fossil energies stood at €38 billion in Germany, 50% more than the EEG—while the EEG is not a state support but a levy. The traditional power companies got their fingers burnt as they underestimated, like many, PV, wind or biopower. E-on, RWE, and Vattenfall lost billions of Euros. After restructuring in 2016, innogy (formerly RWE), E-on and EnBW have become leading investors in the RE field—a new world.

But that is not the end of the story. In March 2018, RWE and E-on announced a new agreement to cut Innogy into pieces again. RWE is to get following those plans all the renewables that were part of Innogy. Not encouraging for the renewables' future with Germany's utilities: RWE is the operator of the country's biggest coal plants; coal is its main interest.

Conventional electricity is no more competitive with solar and its derivates. Full cost of electricity generated from nuclear and fossil power plants stays at some 10 €cents/kWh today. As such, prices can no more be achieved on the spot market; the companies that produce them are in trouble. In auctions of 2017, PV current was offered at 6.58 €cents/kWh: Remember, in 2004, it stood at 45 cents/kWh in Germany. For wind parks, recent auctions gave prices from 5.7 to 4.2 cents/kWh. Even for offshore park, prices came down to 6 cents/kWh.

Germany is globally the only country to shut down all its nuclear power plants. This radical measure follows an age-long popular move against the atom that goes back to the 1970s. A total of 27 GW of nuclear power is being taken off the grid by political decision of the government; by 2017, only 11 GW was left in operation. The country will be totally free of atomic power by 2022. It was a courageous decision taken after a long back and forth in the light of the Fukushima accident in Japan.

Energy experts from all over the world made it clear that both policies, the massive deployment of the renewables and on top the elimination of nuclear power, would end in disaster for the country's energy supply—and perhaps a problem for the world as Germany is one of its leading economic powers. However, what happened gave the lie to the conventional energy thinkers. Contrary to countries such as France and Belgium that kept their large nuclear power parks in place and became electricity importers, namely in the critical winter months, Germany, the leader in solar energy and enemy of nuclear became a big exporter of electricity.

A major problem in Germany's energy system is the role of coal (hard coal and lignite—the brown coal) in electricity generation. Since 2000, both kept their share on the energy market. Only in 2017, Germany decreased a bit its hard coal power, but lignite remained a leader in overall production. Famous are the open-pit mines for lignite in the areas on the left bank of the Rhine, swallowing whole villages and leaving behind pollution and crater landscapes like on the moon. While lignite is still a major domestic energy source, the country's hard coal pits are no more in operation. They were the reason for political conflicts with France about the "Ruhr" and the "Sarre" linked to the two World Wars started between the countries. All pits are closed by now after having swallowed over €100 billion in subsidies and leaving behind big damage in the landscapes and flooded coal tunnels down to 1,000 m deep. Today Germany's hard coal is imported from all over the world.

In the State of North Rhine-Westphalia (NRW) that comprises the Ruhr area, by now three-fourths of all electricity is still produced from coal. Germany still has 134 coal power plants in operation, 51 of them for lignite. Even in Berlin, three coal plants remained in place in 2017 for power generation and district heating, a heritage of communist times.

The big political parties refuse to come forward with a schedule how and when to finish off with all that coal.

And what about the GHG emissions? German governments have a trend to teach the world about fighting the climate change. In fact, Germany's CO_2 emissions of 900 million tonnes a year did not decrease much since 2009. The troublemaker is its wrong coal policy.

By now in 2018, Germany has 64 GW of wind power in operation. Of the total, 7 GW is offshore. In Germany, one counts on average 1,350 equivalent full operational hours per kilowatt and year for wind turbines on land and 2,400 hours for offshore. The reason to go offshore is exactly to have higher operation rates.

The average turbine size in Germany is 2.9 MW. In 2000, it was only 0.7 MW. Dimensions have developed tremendously over the last few years. Most turbine towers now in use are over 120 m high. For offshore, the tower height is somewhat lower. Turbine diameters range from 90 m to over 120 m, for offshore up to 145 m. In total, there are over 28,000 turbines operating in the country.

The biggest offshore park in the region was the one put in operation in the Netherlands in May 2017. It consists of 150 Siemens turbines of capacity 4 MW each. The park cost €2.8 billion and meets 13% of the Netherlands' electricity demand.

The EU Commission has imposed auctions on the RE power markets. From 2019, Germany intends to organise wind power auctions for 2.5 GW of wind power per year. 2017 and 2018 were years with auctions and EEG payments next to each other. In 2017, the EEG payment or the equivalent market bonus stood at some 7.5 cents/kWh. Details depend on the local situation, the wind regime at the place, the hub height and others. The first auction gave kWh prices of down to 4.2 on offer. Also, there was a big success for the proposed community systems. Some offers came in at 0 cents/kWh, meaning that the investor intends to sell his wind electricity on the free market, renouncing any kind of support.

Hydrogen generation from wind power has been tried out in Germany since 2015 to some extent. It is called "power to gas". At the city of Mainz Siemens and Linde operate a hydrogen production plant employing four wind turbines. The total cost of the installation was €13 million, half of it subsidised. In Hamburg, there is a first filling station of hydrogen produced locally. Six local buses and 30 cars with fuel-cell engines are supplied. Toyota is one of the largest promoters of hydrogen-driven cars. Some 3,000 of such cars sold by Toyota were running worldwide in 2017. Alstom in France is even thinking of hydrogen-driven trains.

Germany has some 45 GW of PV power installed in 2018. It generates some 40 TWh of electricity a year in this way. Because of the country's climatic conditions, electricity generation from the overall PV park is concentrated on the summer months. This is in contrast to wind electricity, which comes summer and winter alike. The installation rate of only 2 GW of new PV power per year is poor in comparison of what it used to be 8 years ago. It is small, too, when considering today's tremendous installation rates in China, the United States, Japan or India.

Some 80% of new PV comes ground mounted and the rest building integrated. PV systems over 100 kW get a marketing bonus in the EEG system. At the first auction for large PV parks in 2017—they are also imposed now for PV—the lowest prices came at 6.58 cents/kWh.

The relatively weak interest for building-integrated PV is the more surprising as Germany has passed grid parity for PV electricity with electricity from the grid since 2012 already. **Not** investing in PV on your house in Germany means losing money. Individual customers pay some 28.5 cents/kWh for the electricity they buy from the grid, one of the highest rates in the world. At a price of €1,350 per kW of PV installed on the building, the corresponding electricity price comes some 10 to 12 cents/kWh. The price difference compared with the 28.5 cents/kWh for grid electricity is the net gain. Some auto-consumption is always involved for PV operators on buildings, as only 50% of the electricity generated is qualified for being fed into the grid.

The electricity that is fed into the grid was paid 12.3 cents of FIT for small PV systems on buildings until the end of 2017.

The calculation of the EEG payment and the EEG apportionment is something complicated that the Germans call the "merit order effect". What happens is that on the energy spot market, the electricity price gets all the time lower when more RE electricity is fed into the grids. That has to do with the provision of the EEG that all RE electricity has priority over conventional electricity from coal, natural gas, etc. The RE electricity marginalises the conventional one, which must be sold at a loss, to be sold at all. The more RE electricity is available, the less one needs the conventional one, which must fight for its shear survival on the nets. By mid-2017, the spot market price stood at 3 cents/kWh.

The EEG apportionment is calculated by adding up all EEG disbursements reduced by the income from the spot market where the RE electricity was sold. This is then divided up as EEG apportionment to all electricity clients of the net—with the exclusion of one-third of the clients, the big consumers in industry. They are exempted from the EEG apportionment by the government, which wanted to be nice with its industry: Via the EEG, the industry can buy its electricity at a reduced price on the spot market without having to contribute to the support to the renewables. Germany's big electricity consumers live in a paradise.

In practice, the bill to be paid by the individual clients—it comes out to be some €160 per year for an average client—is the higher the more RE electricity goes to the grid, and the more the spot market price comes down this way. In 2017, in terms of the EEG apportionment, what each grid client, with the exception of big customers, had to pay stood at 6.9 cents/kWh. This represents one-fourth of the clients' total bill of 28.5 cents/kWh mentioned earlier.

In the coming years, the EEG apportionment will go downwards as the RE electricity price will get further reduced and because a growing number of systems will have their 20 years of EEG benefits complete, in particular those from the early days when PV was expensive.

A new trend in the German PV market is to install PV combined with a battery for storage. Electricity storage in buildings is attractive as it allows raising the share of auto-consumption. It is of interest also for older PV installations: After the EEG support comes to an end, auto-consumption is the way to go as the price on the grid, 3 cent/kWh, makes the installation unprofitable. By 2017, 60,000 solar batteries had been installed with the PV plants—more than the 34,000 batteries in the electric cars running in the country. Li-ion or lead-acid batteries are on offer, the latter being the cheaper ones. One must note that the addition of a battery doubles the cost of the PV installation. Batteries with PV installations are extra supported by grants from the KfW bank and tax credits.

Many homes in Germany are rented. Half of the German population lives in rented houses or apartments. Berlin has recently introduced regulations to promote PV installations on such buildings by sharing the EEG benefits between the owner and the tenants.

Promoters of PV in Germany employ various tools for market stimulation. There is a solar cadastre, for instance, Meteonorm developed with WMO, the UN Meteorological Organization, employing satellite data. Renewables.ninja.com is an online tool to estimate solar availability at any location. Data are provided also for passive buildings, wind, etc. Such tools have been used for many cities and states in Germany. In Vienna, Austria, it was used for 240,000 roofs, and in Paris, too. Google Sunroof, the tool for the United States that we have already mentioned earlier, has recently become available in Germany, too. It has all meteorological data for your house, the position of the roof, shadows and seasonal variation. Since May 2017, the tool has been promoted in Germany by E-on and Tedraed: eon-solar.de.

For the monitoring of widely distributed PV plants, the Internet offers, for example, the Internet of Things (IoT) for wireless monitoring.

We have already discussed Germany's leading position on biogas in the special section on it (Section 4.4.3). Just as a reminder, with the EEG 2017, an auction system was introduced that safeguards the German market, which is Europe's leading one in the field. Germany has some 11,000 plants in operation, twice as many as in 2010. The electricity from biogas comes mostly from CHP plants. Some methane derived from biogas goes into the gas networks.

Wood pellets are cheaper in Germany than heating oil and natural gas. The German market started by the turn of the century with 3,000 heating installations for pellets. And like for all other renewables, the market exploded since then. By 2018, 500,000 pellet-heating plants are in operation in Germany.

Germany is also a leader in the field of solar heat collectors. It has 18 million m^2 of solar collectors installed, most of them flat-plate type. Yearly installation rate lies at 100,000 m^2 or so; 2.2 million German houses have a solar hot water (SHW) heater installed.

Germany installs some 80,000 heat pumps per year, too, but it is not a European leader in the field. Systems are offered in a vivid market by Junkers, Bosch, Stiebel Eltron and Viessmann. Air/water systems of 7 to 12 kW capacity driven by electricity or gas are on offer.

Germany, Austria and Switzerland are big promoters of "passive solar heating". Germany has an Institute for passive buildings in the city of Darmstadt. In 1991, Germany's first "passive house" was built there. Freiburg is another centre for the promotion of passive heating—as it is for solar energy in general. In the meantime, many thousands of passive buildings have been built in Germany. Switzerland has its "Minergy buildings". The leader in the field is Austria. Passive houses are mandatory in Vorarlberg since 2007.

In Germany, the standard Energieeinsparverordnung (EnEV) defines the thermal energy demand in a passive house with 15 kWh of heat per m^2 of living area and per year. This corresponds to an equivalent consumption of 1.5 litres of heating oil per m^2 and year. It is only one-tenth of the consumption of "standard houses". The techniques employed comprise solar heating, triple glazing, heat pumps and heat recovery. Passive houses cost 5% to 15% more to build and are financially supported by the KfW bank and regional instruments.

4.4.4 Europe

Europe is well organised in solar and renewable energy matters.

The statistics can be found at Eurostat, which publishes inter alia EU SHARES, Short Assessment of RE Sources. To harmonise RE shares between the different RE technology markets involved and among all the 28 EU member countries, all market volumes are expressed in one single unit, Mtoe (million tonnes of oil equivalent). The drawback is that data become available with 1- or 2-year delay only. Another source of data that I encouraged myself on behalf of the EU Commission in its infancy some 30 years ago is EurObserv'ER, which is published on a yearly basis, too. For details about bio-energy, one can refer to the website of the European Biomass Association (AEBIOM), too.

All the way through the 1980s and 1990s the renewables benefited from political support by the European Parliament. The EU Commission started to move only towards the end of the century. My colleague Arthouros Zervos and I wrote a White Paper that Arthouros was able to convince his Greek compatriot and commissioner to get issued as a communication from the Commission in 1997: "Energy for the Future, Renewable Sources of Energy, White Paper for a Community Strategy and Action Plan" COM (97) 599 final (28/11/1997).

However, it took several more years until the EU Council, which has the final say, started to move as well. That happened, on proposal by the Commission in 2008 and 2009. As a result, Europe has official and mandatory objectives for RE implementation by the year 2020. The EU Directive 2009/28/EC imposes an overall 20% share of RE among Europe's final energy consumption by that time. The directive was adopted when French President Sarkozy and the German chancellor were respectively chairpersons of the EU Council. They took a pro-active stance at the time to get it passed. Chancellor Merkel was cheered with congratulations for it with flowers from the Commission's president.

This overall target of 20% was broken down in 2010 in the NAP, the National RE Action Plans. Each of the 28 EU member countries together with Norway and Iceland has its specific target. Examples are Germany with 18% and France with 23%. Sweden is among the most ambitious at 50% RE penetration by 2020. As it stands by now, most countries will achieve the target they have fixed. However, some big ones such as France or Germany will miss it by a large margin.

Europe's dominating RE source, which comes at 60% of the renewables' total, is **bio-energy**. Since the turn of the century, its contribution has doubled. Next to the traditional wood logs for residential heating emerged a new family of biomass markets. Modern bio-power, biogas, biofuels for transport, wood pellets and municipal waste exploitation gained sizable market shares. Namely, for heating and cooling, bio-energy achieved a penetration of 16% in Europe's final consumption.

It is no surprise that the dominating share of all biomass is solid material, namely wood and other forestry matters. It represents 70% of all bio-energy consumed. Europe benefits from important forest coverage. And that one remains underexploited. Every minute Europe's forests grow by the size of a football field. Removal of wood is mainly directed to the industrial market, and only for some 20% to energy.

In Europe, money grows on trees. Total turnover of bio-energy exceeds €60 billion per year; 500,000 jobs are involved, more than for wind power or PV.

The main consumers of bio-energy in Europe are Germany, France, Sweden, Italy, Finland, Poland and Spain. The same countries are Europe's leaders for heat consumption from biomass. Sweden, Finland and Denmark have an important district heating market.

Most of the world's **wood pellets** are consumed in Europe. In 2015, the European pellet market crossed 20 million tonnes—by the year 2000, it just did not exist. The top producing countries for pellets today are Germany, Sweden, Latvia, France and Portugal.

The **United Kingdom** plays a particular role in the pellet market: It consumes one-fourth of Europe's pellets, all of them being imported from North America, while the rest of Europe relies on its own production; and all of those pellets in Britain serve the power market, while in the other countries they are used for heating. We have already mentioned the Drax coal plant in the United Kingdom. That one was converted for 70% to **co-firing** with pellets. Drax produces 20% of Britain's overall renewable electricity.

When the Brexit becomes effective, the rest of Europe will change its pellet markets quite a bit: It will greatly reduce imports of pellets and their use for co-firing, too.

Well, not completely. **Copenhagen** in Denmark projects to replace by 2019 a 600 MW coal plant with a CHP plant fired with imported pellets. It is expected to cost €150 million to build.

The other newcomer in bio-energy since the turn of the century is **biogas**. We reported about it in preceding sections, in particular that on Germany. In Europe's energy consumption, biogas stands for some 16 Mtoe. This is actually a lot more than the 21 million tonnes of wood pellets considering that energy-wise 1 kg of oil or biogas is equivalent to 3 kg of wood. Contrary to pellets, which are mostly used for heating, biogas is in general burnt in CHP plants providing next to the heat high-value electricity.

A special case is **Sweden**, where biogas is mostly used for heating only. The other particularity is that one-fourth of the total is used for transport.

In the rest of Europe, biogas is not much employed for transport. This place was taken in the year 2016 by **biodiesel**, with 10.9 Mtoe, and by **Bio-ethanol**, with 2.6 Mtoe. Together with the 5.8 billion litres of ethanol, the industry produces 5.9 million tonnes of co-products, mostly high-protein animal feed. The **biofuels** came 5.3% in Europe's transport market in 2016; biodiesel's share is 5.8% and that of bio-ethanol 3.3%. The use of pure vegetable oil is negligible. In the RE Directive from 2009, a 10% incorporation rate of biofuels in European petrol by 2020 had been decided. That target was taken down in 2015 to only 7% in the so-called Indirect Land Use Change (ILUC) reform.

Sweden uses next to some biogas a lot of liquid biofuels. It is Europe's champion with 19% incorporation, mostly diesel. The other European leaders for biofuel's use are France, Germany, the United Kingdom and Spain.

In energy importance then, the 13.5 Mtoe of biofuel liquids come second after the 16 Mtoe of biogas and before the 7 Mtoe of the solid pellets.

The energy derived from **municipal waste** represents some 7% of the total of biomass—comparable to the share of pellets.

Coming back to heating, we have to address the **heat pump market**. There are some 30 million heat pumps in operation in the EU. Every year, about 2.5 millions of them are newly installed. They count for 9 Mtoe of RE heat produced. More than half of all heat pumps are installed in Italy. They serve for heating and for cooling, too. The other lead countries for heat pumps in Europe are Spain, France, Portugal and Sweden. Virtually all of them are aero-thermal and use air-to-air technology. In Germany, heat pumps are not very popular. Those in use employ another cycle, the air–water. Geothermal heat pumps have a European market of less than 100,000 units only.

Sweden derives 10% of its heating energy for buildings from heat pumps; 55% is provided by biomass leading to a record of 65% of RE heat in the whole building sector.

And there is **heating with solar collectors**. In 2016, a total of 51 million m^2 of solar heat collectors were in operation in Europe. The top countries with the largest parks are Germany, Austria, Greece, Italy, Spain, France, Poland and Denmark. The order of countries for new installations is a bit different, as Poland, Greece and Denmark come directly behind Germany, which keeps the lead with almost 1 million m^2 of collectors installed in 2016. Most solar collectors in use in Europe are the flat-plate type.

The role of Turkey in this field is the most impressive. Turkey is the world's second in terms of new installations after China, ahead of the United States and India, which come third and fourth, respectively, and ahead of all the EU countries.

Wind power and PV are the big success stories of Europe. They started their conquest of the energy's mainstream at the turn of the century.

In 2017, the EU had 170 GW of **wind power** installed; in 2000, it had been only 17.6 GW. It comes now second after China and ahead of the United States. For new installations, the same order prevails; first China, second the EU and third the United States.

At the end of 2017, Europe's offshore wind capacity stood at 13.7 GW.

The EU produced 336 TWh from the 170 GW of wind power it had installed up to the end of 2017. Its capacity factor or equivalent full time of operation over the year was then 20%. Britain has Europe's highest at 25% followed by Denmark, Spain, France, and Italy with 18%. Germany has the most unfavourable wind regime with 16% only.

However, Germany puts a brave face by still maintaining its leadership in European wind power installation and production. End 2017, it had, as mentioned previously, over 55 GW of capacity installed, generating 101 TWh. In capacity size and production, too, Germany was followed by Spain, the United Kingdom, France, Italy and Denmark. Spain had zero wind power growth in 2016 and France the highest after Germany.

The European **PV capacity** passed the 100 GW mark in 2016. Until that year, the EU was the world's number 1 before being overtaken by China in 2017. With a production of 105 TWh in 2016, Europe's PV capacity factor was just over 10%. It is the same for Germany, Britain and Belgium, where it comes under the European average of 10%. For Italy, it is 11% high, and for France 12%. The best solar climate has Spain with 19%—at least that is what Spain claims.

Germany has the highest PV capacity installed followed by Italy, which has 50% less. Then comes the United Kingdom, before France, the Netherlands, Spain and Belgium.

The PV markets in Europe exploded in two waves—not counting at this stage the pilot phase before the year 2000. The first one was led by Germany and Spain between 2000 and 2010–2011. The second one started only lately in 2011, led by Britain and Italy. Between 2011 and 2015, the PV capacity in Britain increased by a factor of 300 (three hundred!)—it had switched to FIT promotion. Still in 2015, new installations of PV in Europe were led by Britain with 3.76 GW ahead of Germany with 1.5 GW, and France, the Netherlands, Italy and Belgium, which followed.

This was again an example that **PV and wind markets alike are not following only local climatic conditions but policies**.

In conclusion, the European Union produced overall some 16% for its final energy consumption in 2015 from RE sources. That share doubled from 2004; at that time, it had been provided by hydro and conventional wood logs for heating.

EU electricity share from RE stood at 29% of the final electricity consumption in 2015. Hydro and wind electricity made up each a third of the total, followed by biopower, PV, etc. Wind power and PV have the highest growth rates. It is interesting to note that more than half of all RE electricity comes in the form of hydro and bio-energy, which are continuously available, ready to complement the intermittent sources solar PV and wind electricity. This is a further encouragement towards the development of a "100% RE Energy Europe".

EU **heat generation** from RE stood at 19%. The main constituents were 80% solid biomass, 9% heat pumps, 3.5% biogas and 2.2% solar thermal collectors.

Total yearly turnover of European RE generation in 2015 was €153 billion, led by Germany, France and the United Kingdom. Forty percent of the turnover stood for bio-energy, followed by wind power, heat pumps and PV.

In Europe, 1.14 million **jobs** are provided by the RE implementation, most of them in Germany with 322,000 followed by France, Britain, Italy, Spain, Sweden and others.

A last word about **Paris**, France, the city where I live. It is a mega-polis, populous, a hotspot in international tourism and the venue for the 2024 Summer Olympics.

Paris has several well-organised public transport networks. Most Parisians have given up driving their own car; only 1% of them take the car to go to work. Two key highways crossing the centre of the city along the river Seine have been turned into pedestrian boulevards. Two bus lines are running on biogas.

On almost every street corner, the city has established centres for the location of bikes or electric cars. Right in the centre, a new filling station for hydrogen fuel has been started. Even bicycles running on hydrogen are being tested.

Contrary to neighbouring Brussels, not much PV integration can be seen as yet on Parisian buildings, but there is hope. France has introduced a law in 2015 making PV or green roof integration obligatory on new commercial buildings.

4.4.5 Japan

The country is a great pioneer in the promotion of PV. PV has traditional priority over all other RE technologies in the country. Over 90% of Japan's renewable energy comes as PV.

After the first oil-price shock, Japan started a "Sunshine Project" in 1974, which was followed in 1993 by a "New Sunshine Programme". A year later, the government decided the "Basic Guidelines for New Energy Introduction". It called for the installation of 400 MW by 2000, equivalent to 100,000 PV roofs, and 4.6 GW by 2010.

Japan gave an investment credit of 50% when a PV system cost €10 per Watt, which was reduced to one-third of the cost since 1997 for private homes. However, this financial support was insufficient to stimulate market growth. By 2000, only 70 MW of PV were installed in the country that year. By 2009, it had not increased too much when it remained at 400 MW only. In the meantime, Germany, Spain and all those countries that had adopted the FIT became PV importers and Japanese industry took advantage of it. In 2008, Japan produced 1.22 GW of PV modules and exported 80% of them.

The change came in 2012, when like China, Britain and many other countries, Japan introduced the FIT. As a result, markets exploded and Japan turned from a PV module exporter to an importer. The other effect of the FIT introduction was that large-scale PV plants took the helm over the residential systems, which had been the dominant markets before. Japan adopted a very generous tariff. Even the revised one for the fiscal year 2017 provides 23.9 $cents per kWh for systems up to 10 kW.

Shortly after the introduction of the first FIT in Japan, the yearly installation reached an absolute maximum of 9.2 GW in

2014, mostly for utility-scale and commercial systems. Then it decreased year by year again coming down to some 6 GW in 2017. Cumulative PV capacity in Japan at close to 50 GW comes third globally after China, and the United States, and ahead of Germany, which it passed in 2016.

4.4.6 India

If one believes the official figures, an energy revolution is going on in this country. India is a traditional coal country. That is right now about to change.

Just in May 2017, 14 GW of new coal was cancelled. In financial year 2015–2016, India still installed 23 GW of thermal power and only 7 GW of renewable power. A year later, for the first time both came at the same level: Not more coal was installed than the 10.9 GW of renewable power, 5.4 GW of new wind, and 5.5 GW of new PV. Electricity from imported coal is no more competitive with PV. In 2017–2018, a record power of 9.5 GW of PV has been newly installed, bringing the total to some 20 GW of capacity in addition to the 35 GW of wind power. In 2017, India passed for the first time Japan in terms of annual PV installation.

Wind power was an early starter in India, too. From just 6 GW in 2005 it increased to some 40 GW during 2018. Almost 10% of India's electricity demand is met today by wind power. Over 400,000 people are employed in the sector. In 2017, prices down to 4 $cents/kWh have been achieved in auctions.

After tendering in 2017, PV electricity prices as well came down to 4.4 cents/kWh. The installation price can be below $1,000/kW per PV plant. That is helped by lower interest rates. Most of the PV plants are utility scale and only some 15% get on buildings. One-fourth of the electricity is sold to Delhi railways. Every train station will have PV, it is said.

The government has declared that by 2022, 100 GW of total PV capacity will have to be installed.

However, PV in India seems to be a medal with two unequal sides. It is claimed that some 230 million people still have no access to reliable electricity although the government claims the contrary is true, and most villages were electrified by 2017—an interesting development to follow.

Let's also mention that India is busy developing its water pumping infrastructure with PV. By 2017, over 30,000 PV water pumps are said to be running. A million are promised by 2021.

4.4.7 Brazil, Latin America

Brazil is a model country for renewable energy implementation and use. It is ahead of all big nations and regions of the world, the United States, Europe—even beating by a mile Germany and China. Today, Brazil already covers 83% of its electricity demand by the renewables: 61% from hydro, 9% from bioelectricity and 7% from wind power. The 1% of atomic power is particularly ridiculous: I remember all the efforts deployed by the United States and Europe to get Brazil committed to nuclear power in the 1970s. The country was one of the few not to listen for very long to the seducing atomic sirens with their money pockets.

What was missing so far in the renewable bouquet was PV. I had myself already promoted PV there in the 1970s, but I was not convincing enough. Brazil's PV market started slowly only in 2015. At last, by 2018 was completed at Minas Gerais a 400 MW PV plant, Latin America's biggest. The prime contractor was EDF. Its PV modules were provided by Canadian Solar. It was imposed that 60% of the plant's value had to be generated in the country. This was used as an excuse to justify the €1,300 per kW installed, which is a bit on the higher side of prices in today's global markets.

We mentioned earlier Brazil's pioneering role carried through until this day against all odds for bio-ethanol promotion in the transport sector. Brazil has a long tradition, too, for employing charcoal from short-rotation forestry—instead of mineral coal—in the steel industry.

In the rest of Latin America, another whole family of utility-scale PV plants came into operation recently. A 246 MW plant of Acciona and another 147 GW plant by EDF are in operation in the Atacama Desert in Chile. A 160 MW plant from ENEL Green Power is at Finis Terrae at Antofagasta, also in Chile. Two more of similar size from First Solar and SOPOSA are running in Honduras and Chile. A 101 MW PV plant is in operation in El Salvador since May 2017. All those plants cost on average $1,500/kW to build.

There are smaller PV systems, too. Nicaragua is supported by the United Nations Environment Programme (UNEP) and the International Monetary Fund (IMF) to make it by 2020 "Renewable Energy" country with ample hydro, biomass, wind power, geothermal and solar.

In Bolivia, President Morales inaugurated in 2017 a 5.2 MW PV plant, and 93 MW of wind power was installed with the help of Denmark.

In a later chapter on Development Aid, we will come back to Central America, where the EU has helped to install PV/wind systems for Internet connection in 600 schools.

Part 3

Understanding Nature, Creating Know-How

Chapter 5

Splitting the Atom and Creating Solar Technology

5.1 Quantum Physics and Understanding the Atom

Max Planck is the founder of quantum mechanics. He presented his findings in Berlin in 1900. From the interpretation of the spectrum of light, he concluded that light must be composed of quanta. He introduced Planck's constant *h* and the notion of energy quanta. It is the basis for understanding the world of microphysics, the world of atomic dimension.

His friend Albert Einstein played a big role to encourage him to believe in his findings—as Planck was reluctant to believe in what he had discovered himself. It was then Einstein in 1905 who explained that light is composed of quanta—although the term photon stems from 1926 only. And Einstein got the Nobel Prize for interpreting the photo-effect. The photo-effect is often mixed up with the photovoltaic effect, but PV is different from the simple emission of electrons by light as we will see here below.

Planck was also the organiser of the first Solvay meeting in Brussels in 1911 where all the world's leading scientists in physics of the time met, Planck, Einstein, Rutherford... Figure 5.1 shows a

The Triumph of the Sun: The Energy of the New Century
Wolfgang Palz

ISBN 978-981-4800-06-8 (Hardcover), 978-0-429-48864-1 (eBook)
www.panstanford.com

copy of the photo of the meeting taken by the author. It hangs till this day at the reception desk of the Metropole Hotel in Brussels, where the meeting took place.

Figure 5.1 Brussels Solvay meeting (1911) of the world's leading physicists. (Standing first on the right) Albert Einstein, (third from the right) Ernest Rutherford and (standing second from the left) Max Planck.

New Zealand-born Ernest Rutherford at the University of Manchester in England was the one who explained for the first time that all atoms are structured along a planetary geometry, with the nucleus in the centre and the electrons on shells around—an obvious difference being also that our Sun's planets move approximately in one plane, while electrons move on shells. All space in between the tiny nucleus and the shells is virtually empty. Rutherford concluded his findings from the analysis of the spectral lines of hydrogen in 1911. Until then one had thought that the interior of the atom is composed of a soup of positive and electric charges, the plum pudding model. Here we encounter again the enigmatic and colossal force keeping all positive protons together in the nucleus. We discussed it in the context of the interpretation of the Sun's energy. It is at the origin of those energies liberated when an atomic bomb explodes.

In 1911, also entered the stage Niels Bohr from Copenhagen. He spent some time as Rutherford's assistant in Manchester before returning to his hometown in Denmark. In 1913, he published Bohr's equation developing further the planetary model of Rutherford. He combined the planetary model with Planck's energy quanta. Bohr found that the electrons evolve on stationary orbits that stand for well-defined energies. No other orbits are allowed. But electrons can pass between different orbits and only if there is space: On each orbit, the number of possible electrons is well defined. Inner orbits are full. Chemistry is nothing else than the play between the reaction of different atoms and the electrons on their outer shells.

5.2 From Quantum Physics to Nuclear and Semiconductors

After World War I, the 1920s and 1930s were a fascinating time of further development of the knowledge of the atom and quantum mechanics. Dozens of extraordinary brilliant scientists were involved. Perhaps because the father of it, Max Planck, had started the work from Berlin, Germany was the land where much activity was centred—until the Nazis took power in 1933.

The town of Göttingen in central Germany became the world's capital of mathematics and theoretical physics. Professor Max Born had here the university chair devoted to quantum mechanics and solid-state physics. Pascual Jordan, Viktor Weisskopf, Wolfgang Pauli, Werner Heisenberg and many others worked with him in Göttingen. Also Robert Oppenheimer, Edward Teller, John von Neumann and Enrico Fermi prepared with him their thesis or were his assistants, those who built later at Los Alamos the world's first atomic uranium and hydrogen bombs. The cemetery of that little town still has the tombs of eight Nobel Prize winners, Max Born, Max Planck, Otto Hahn, Max von Laue, Walter Nernst, and others.

Born and Einstein were good friends. Until Einstein's death in 1955—Born was back again in Germany—they exchanged 117 letters, it is said.

Another leader was Arnold Sommerfeld. He made mathematical contributions to the special theory of relativity of his friend Einstein. When he had a university chair in Munich, Werner Heisenberg was one of his assistants. Among his many students were Linus Pauling, Max von Laue and Hans Bethe, a wide-ranging scientist who also became one of the leaders at Los Alamos a few years later.

Later Werner Heisenberg himself became a leader in his own right in Leipzig. Among his students were Pascual Jordan, Carl Friedrich von Weizsäcker, Felix Bloch and Edward Teller. This "hydrogen bomb" Teller had previously graduated in Karlsruhe, the same University where I passed my thesis some 30 years after Teller.

Other physicists of those exciting times were Erwin Schrödinger, who succeeded Max Planck in Berlin, and Liese Meitner, closely related to Otto Hahn.

And a kind of godfather was Niels Bohr, sitting in Copenhagen. Sommerfeld, Heisenberg, Pauli and many others have been his assistants.

All these scientists knew each other very well, exchanged ideas, often shared an apartment and worked in the same office. They spent much of their free time together. Many liked hiking. In a book (in German) about the "bomb builders," published by dtv in 2013, Alex Capus reports how Heisenberg, Niels Bohr, his son Christian, C. F. von Weizsäcker and Felix Bloch went for the last time skiing together in the Bavarian Alps. That was in 1933.

Many of all these scientists were Jews or had a Jewish family and saw no other issue when the Nazis took power in Germany than to emigrate. Einstein is a well-known example. Many went to the United States and ended up in Los Alamos: Hans Bethe, Felix Bloch, John von Neumann, Enrico Fermi and Edward Teller. Robert Oppenheimer came from a Jewish family of German origin but was born in the United States. Max Born went to Britain but came back to Germany after the war. Lise Meitner escaped to Sweden and so did Niels Bohr when the Nazis occupied Denmark.

5.2.1 The Way towards Nuclear Fission

A breakthrough in fission research occurred in 1932. In that year, James Chadwick, an assistant of Rutherford in Manchester, discovered the existence of the neutron. One knew well that charged particle like the protons have no chance to access the atomic nucleuses charged positive as well—the repulsive force is insurmountable—but a neutral particle can.

One of the first to understand the new opportunities was the Hungary-born Leo Szilard. Right away in 1933, he conceived a nuclear chain reaction and took a patent on it with Enrico Fermi. On his part, Werner Heisenberg was the first to develop a realistic model of the atomic nucleus composed of protons and neutrons.

Otto Hahn succeeded in splitting the atom and proving the experimental evidence in 1938. As we have reported previously, it was Frédéric Joliot in France, the son-in-law of Madame Curie, who succeeded first in proving experimentally a chain reaction in uranium. And he also took a patent on a bomb...

5.2.2 The Origin of Solid-State Physics

Thus far, we have discussed only the isolated atom. What happens then in condensed matter, in crystals with many atoms aligned in order? Felix Bloch, the Swiss who also went to Los Alamos—we have to report on his time there later—and eventually ended up as the first director of CERN in Geneva, addressed this problem in 1928. Referring to the important "exclusion principle" of Wolfgang Pauli that introduces different spins for the electrons in an atom, Bloch developed in Leipzig in his thesis with Heisenberg, Bloch's theorem. The theorem explains that the discrete energy levels of the electrons in a single atom become broadened into bands in solid matter through interaction.

Shortly after, in 1930, **Kronig**, a German American, published in London his quantum mechanical model, the Kronig–Penney model. Kronig lived in Delft, the Netherlands, and published *Band Spectra and Molecular Structure* with Cambridge University Press. He was connected with Bohr, Heisenberg, Pauli, ...

Semiconductors, be it for ICs, LED lamps, optical screens, or PV, obey this model. Electrons are confined not on a single energy level but in bands, the valence band being the outer one. The valence band is separated from the next outer band by an energy gap where electrons are not allowed following quantum mechanics.

Electrons can be raised to the empty conduction band by either doping or optical excitation. Both effects are involved in the PV of a solar cell. Phosphorous "impurities" can be implanted in the forbidden gap of crystalline silicon. The phosphorous atom has one more electron than silicon on its outer shell. That electron is released to the conduction band. And in this way, that part of the crystal becomes n-type. Correspondingly a p-type area is created by doping, for instance, with minute amounts of boron. The boron atom has one electron less than silicon on its outer shell. In the silicon crystal, this deficiency is compensated by an electron jumping up from its valence band, leaving behind a movable hole. As a result of all this, a barrier layer is created between the n- and p-type areas of the silicon crystal, called a diode. As mentioned earlier, such diodes are the basis of the whole semiconductor world. Moreover, once a diode is illuminated, the photons raise electrons from the valence band, leaving behind the holes. This is the PV effect. The electron–hole pairs created by the light are separated by the voltage drop in the barrier layer and become the PV current.

5.2.3 The Atomic Bomb and Nuclear Reactors

After the fission of uranium had been demonstrated, began in America the Manhattan Project. It was one of the most extraordinary technological and industrial endeavours mankind has ever achieved—and ended with the death of 250,000 people in just two days, also a record.

The Manhattan project was part of WWII. It was directed from the same people who had fled Germany and its objective was to destroy that same country by some super massive bombs the world has never seen before.

It started with a letter sent in 1939 by Einstein, Szilard, and Wigner to President Roosevelt asking to develop the atomic bomb. Roosevelt approved in late 1941. In June and July 1942, Oppenheimer and Fermi convened two meetings at Chicago and Berkley, California, to start the project. All the other guys we have seen before in Germany also participated: Hans Bethe, Edward Teller, von Neumann, Felix Bloch... Teller pushed right away for the development of a hydrogen bomb, but he succeeded with that idea only later. He also expressed the fear that with a super bomb, Earth's whole atmosphere could be ignited through fusion, but the intelligent Bethe calculated that this was impossible.

On 2 December 1942, Fermi got the world's first atomic reactor with uranium and graphite critical in Chicago. This was a challenge. In fact, one did not know much about criticality in those early days. If one comes close to self-sustained reaction and it becomes critical, the whole thing can explode by the snowball effect. This is what happens in the bomb of concentrated uranium-235 when there is a "prompt critical" reaction. On the other hand, a nuclear pile with uranium-235 and 238 is constructed in such a way that the nuclear reaction is self-sustained by a mixture of "prompt" neutrons originating directly from the fission of uranium-235 and slowly released neutrons from isotopes that absorb some neutrons and release them again with delay. Unlike in the bombs, in nuclear reactors, this effect of delayed neutrons provides some range of criticality that is broader than 1. Only if one goes beyond that range, the reactor explodes, as happened at Tchernobyl.

Commercial reactors today use 3% to 5% of fissile uranium-235 together with uranium-238 that is not directly fissile but converted to plutonium-239. The fast neutrons emitted from the split uranium-235 atoms are slowed down by a moderator, otherwise they are absorbed by the uranium-239, a process that is undesired in a commercial reactor, even though being not totally avoidable. We mentioned previously that a 1,000 MW reactor of today produces 290 kg of plutonium every year.

Another advantage of working with slow neutrons is that the probability of capturing them for further splitting the U-235 and producing energy is higher than it is for the fast neutrons. Most reactors today employ water as a moderator, some use solid graphite and just a few use heavy water. Needless to say, bombs have no moderator and directly use the fast neutrons for proliferation.

In nuclear "breeder" reactors, one uses the effect of conversion of U-238 into plutonium by the fast neutrons. Producing energy and new fuel at the same time. There was a time when one had put much hope on this technology. The highlight was the French Superphénix reactor. However, in France and worldwide, all efforts on this technology have been given up, except in

Russia. Near the Ural, the BN-600 and BN-800 are in operation and an even larger one the BN-1200 is being planned.

In the Manhattan Project, most effort went into the production of nuclear fuels. U-235 was produced from natural uranium by three different methods: electromagnetic, gaseous and thermal diffusion. A fourth one, centrifugal separation from U-238, did not work then. The other fuel was plutonium. It was a new element that had only been discovered in 1940 in California. For the Manhattan Project, it was produced from U-238 in a water-cooled reactor at Hanford.

Los Alamos was created out of nothing by Oppenheimer in late 1942. The first bombs were designed and built there. All the experts mentioned earlier participated and moved to Los Alamos in a total secrecy. Bloch, who was in charge of the bomb's design quit in 1943 to the vexation of Oppenheimer. It became clear to the Swiss then that Hitler was going to lose the war anyway. No need to drop atomic bombs on Germany anymore. Not so nice to think that he joined the Manhattan Project only with the prospect to destroy Germany with those super bombs.

By July 1945, everything was in place in Los Alamos. Enough fuel had arrived and so was produced the uranium bomb that was dropped in August on Hiroshima and the plutonium bomb that was dropped from an airplane on Nagasaki.

Szilard and Einstein wrote to the president urging him not to bomb the Japanese cities. But Oppenheimer, Fermi, Compton, and Lawrence recommended the bombing. Fermi is quoted: "Don't bother me with your conscious scruples. After all, the thing is superb physics". The letter demanding not to go ahead never reached the president. Roosevelt had already died and Truman was in favour of doing it, the bombings.

In my younger years, I went to see the mock-ups of the two bombs thrown on Japan that are shown at Albuquerque, New Mexico, in the United States. I also went to see Hiroshima.

As mentioned earlier, it was Robert Jung, a Jew from Berlin who had escaped in the war to Switzerland, who was the first outsider to get access to Los Alamos in 1949. After discussing with Oppenheimer there, he got away with a bad opinion about him. He published the book *Brighter than a Thousand Suns* to report on the Manhattan Project. The leader of all what had happened, Robert Oppenheimer, got under attack from 1953 onward in the McCarthy trials. He was not much supported by his former colleagues. He said about himself: "In Los Alamos we did the job of the devil".

5.2.4 A New Semiconductor World and PV

The first practical p-n junction on a semiconductor—it was a silicon crystal—was demonstrated in 1939 at Bell Labs, Murray Hill, New Jersey. All semiconductor diodes we use today, the LEDs, or the laser diodes, have their roots in this device. **This was also the birth hour of PV in silicon**; when that diode was illuminated by the dish lamp, it showed a PV effect.

Russel Ohl, who performed the experiment, had the right intuition to interpret the result by barrier formation through doping on different sides of the crystal. It were actually different impurities on each side and the doping had been done accidentally. Later during the war, Bell Labs employed germanium diodes in radar units.

The concept of a tri-polar transistor had been invented much earlier by **Julius Edgar Lilienfeld** in Leipzig. He got it patented in Canada in 1925. It was the concept of the field-effect transistor MOSFET that is nowadays a fundamental element in integrated circuits. By 1934, an **Oskar Heil** in Göttingen introduced another patent on the idea of the transistor in Britain.

Also at Bell, **John Bardeen** created a new branch of quantum mechanics to understand electron mobility in crystals. He built the first working transistor in December 1947. As he worked together with **Walter Brattain**, who did the experiments, and **William Shockley**, who was the boss, all three together took the patent on the "point-contact" transistor. The FET could not be patented because of the prior patent of Lilienfeld. In 1948, Shockley took alone the patent on the first bi-polar transistor, an n-p-n junction transistor on germanium. He took the patent without his co-workers because the relations with his colleagues were poisoned and full of jealousy.

The Bell group had initially published nothing about its inventions; so in Europe another group simultaneously developed the same transistor. It was the German **Herbert Mataré** and co-workers, established in the Paris region, who invented in 1948 what they called the transistron. Mataré was an expert.

During the war, he had worked on silicon in Germany. Later, he was the first to commercialise diodes and transistors in Germany and to sell the world's first **transistor radio**, a year before the Americans did.

In his late days, I had a chance to meet Mataré in Aachen, his hometown.

However, all this work on poly-crystalline germanium had not much impact on the semiconductor and informatics world that followed. The real thing that was invented a bit later was the **silicon transistor**. It was the merit of **Gordon Teal at Texas Instruments**. Teal revealed his achievement, the first commercial Si transistor, in April 1954 to the public. Teal had previously worked at Bell Labs, too. And it was actually there at Bell that Moris Tanenbaum built the first transistor device on silicon using a technology used some months earlier by Teal when working there.

The first thing Teal found necessary was to work on highly pure mono-crystals. He grew silicon crystals by the method that the Pole Czochralski had already invented in 1915. The high-purity, semiconductor-grade silicon material was provided by Dupont. On such silicon crystals, Gordon then developed the n-p-n bipolar transistor. It was immediately commercialised with enormous success and made Texas Instruments known worldwide.

In the same year, 1954, and at the same place, Bell Labs., the first commercial silicon solar cell was developed—by other people. More on this later.

Integrated Circuits (ICs) were also invented at Texas Instruments. It was the merit of **Jack Kilby**, who realised them in 1958 with colleagues. He received the Nobel Prize for it and the praise of the US president.

ICs on silicon chips: another success story of modern times. In 1971, there were 2,300 transistors on one chip. In 2014, that number reached 2.6 billion. Silicon ICs are in use today about everywhere in modern electronics, personal computers, mobile phones, data centres and telecom services.

In 2015, the global spending on information technology (IT) devices reached $725 billion, almost 10 times more than the global PV market that year.

The same semiconductor-grade silicon is employed for ICs and PV solar cells alike. We attempted in the early days to develop "solar-grade" silicon with the idea that for solar cells the purity requirements for the material were less stringent and production cost might be lower. However, it turned out that the global mass production of semiconductor-grade silicon increased to such an extent that prices collapsed and became low enough also for very cheap solar cell and module production. More on this later, too.

IC chips are some five times thicker than the silicon wafers employed for commercial solar cell manufacturing.

Area-wise, the world employed in 2015 a surface of 6.7 million m^2 of silicon chips. If one had made on that area solar cells instead, the PV power achieved then would have been equivalent to some 1,000 MW or 1 GW. This compares with an actual installation rate of 50 GW of real silicon solar modules that year. As a matter of fact, the world uses since 2006 more semiconductor-grade silicon for PV than it does for IC chips.

Optoelectronics is another semiconductor sector that became important more recently, in particular with the emergence of LEDs for general use. The p-n diodes of silicon do not emit light. So one had to go for other materials. One possibility is GaAs. Texas Instruments got a first US patent with it on a light-emitting diode in 1962. But that diode emitted only infrared light. What followed was a rush with always more types of semiconductor diodes for all colours of the visible spectrum. The winners were **Nakamura**, **Akasaki**, and **Amano** in Japan, who showed the first white light-emitting diode in 1994 and later received the Nobel Prize in physics for it.

The early LEDs had two important drawbacks. The light intensity was insufficient for practical purposes and they were far too expensive. In the last few years, both problems were solved. Up to 300 lumen per Watt have been reached and prices cut to a tenth. Philips, a world leader in the business, achieved a turnover of €7.5 billion with its LED lamps in 2016 and expects 25 billion by 2023. This being said, LED has not yet achieved a monopoly position for lighting as halogen lamps and low-pressure neon lamps defend for the moment their important market shares in lighting.

Many of the successful LEDs employ compounds of Ga, As, In, P and N. The material is heavily doped.

LED has made important inroads into the world's optical display market where billions of computers, smartphones and

TV sets are sold. LED is employed in combination with liquid crystal (LC) technology. OLED and QLED are promoted for TV screens, but at this moment actual LED pixels allowing brilliant pictures instead of being used as backlighting of LC displays are yet in the trial stage.

Chapter 6

Photovoltaics

6.1 The World's Global PV Markets: How They Exploded

Previously, a lot of details have already been given on PV's adventurous route since it made its appearance on the world's power markets. Here is a summary of what happened.

The global PV markets came in three waves. The first one stretched from the invention of the silicon solar cell until 2000 when globally installed capacity reached 1 GW.

The second wave went over 25 GW, reached in 2009, all the way up to 100 GW end 2012.

The third wave since 2013 brought even more accelerating markets with 450 GW achieved in 2018.

In the first 20 years after the invention of a practical solar cell, the world achieved by 1973 a meagre 1 MW of terrestrial capacity installed, led by the United States. By 2000, that market had cumulated to 1,000 MW, or 1 GW, by trial and error. That market was led by Europe, the United States and Japan. The second wave came with the beginning of the new century when Germany started the FIT incentive. Accordingly, Europe directed the world to a record PV capacity of 100 GW by 2012. The top countries were Germany and Spain. The third wave looks rather

The Triumph of the Sun: The Energy of the New Century
Wolfgang Palz

ISBN 978-981-4800-06-8 (Hardcover), 978-0-429-48864-1 (eBook)
www.panstanford.com

like a tsunami. It started when China, Japan, the United Kingdom and others adopted the FIT system in 2013 and their markets exploded. Some 450 GW of PV power in total have been installed by 2018. At the same time, Europe lost its leading role since that crucial year (2013) and was completely sidelined. Since 2016, the world has become used to a cruise road of 70 to 100 GW newly installed PV power every year. The top nations in this rush are China, the United States, India and Japan.

Since the beginning, global markets were dominated by silicon solar cells and modules, even though a niche was always left to the alternative technologies that have been always around.

6.2 State of the Art of Today

Today's global PV market is dominated by silicon solar cell modules, which occupy an overall share of 94% of the total, 71% for multi-crystalline silicon solar cells and 23% for single-crystal cells (figures for 2016, after RTS Corp, Tokyo, Japan). In general, multi-crystalline modules look blue and the single-crystal ones black. Thin-film modules of CdTe with a 4% market share and those of CIGS with 1.5% play a marginal role. China and Taiwan produce the lion's share of 73% of the PV world market of modules.

On the international spot market, **silicon module prices** came all the way down from $1.85/Watt in 2010 to **30 cents/Watt today**. All parts of the value chain were concerned by this tremendous cost decrease. In the period since 2010, the **silicon feedstock price** came from $80/kg to $12/kg, the **solar cell price** stands by now at $1.14 for a standard 6 by 6 inch cell, 5 times less than that in 2010, and the cost of the silicon wafers decreased similarly.

The **classical crystalline silicon solar cell** is made of a p-n junction, a diode. Light absorption on the top surface is improved by an anti-reflective coating after grooving it by etching. On the backside, the cell has an aluminium contact and a back-surface field (BSF). Such cells have typically an efficiency of 20%.

Currently, **passive emitter rear cell** (PERC) and **hetero-junction with intrinsic thin layer** (HIT) cells have come to the forefront. They are upgraded versions of the classical cell.

Figure 6.1 A multi-crystalline silicon solar cell (picture by the author).

The **PERC** has in addition a dielectric passivation layer on the backside that reflects the non-absorbed light back into the mono-crystalline silicon to give it a second chance for absorption. The Swiss Meyer Burger is a leading company for the needed coating equipment. PERCs have a slightly higher efficiency up to 22%.

The **HIT cell** is produced by Panasonic in Japan. It consists of an n-type silicon crystal that is on the top surface covered with an extremely thin layer of amorphous silicon, just 100 atomic layers thick. The contact layer is intrinsic, or non-doped, covered by a p-type layer of the same fine amorphous silicon. Together, the barrier, a hetero-junction diode, is formed between the n-type crystalline silicon and the p-type amorphous silicon separated by an intrinsic part. Panasonic's best efficiency achieved so far is 25.6% on a cell and 23.8% on a complete module. Those are obviously laboratory items. Modules for market sale come with some 20% efficiency. Panasonic is known to have a co-operation agreement with Tesla, the electric car manufacturer in the United States.

Still another technology goes beyond the HIT of Panasonic. It does not have a name on its own. It is very sophisticated and perhaps the most interesting of them all. It is **Kaneka's silicon crystalline "hetero-junction, back-contact type" cell**. To form the hetero-junction, the process employs a very thin intrinsic and p-type amorphous silicon layer like the HIT cell, but here it is put on the backside of the cell, the one not illuminated. Another difference is that both contacts, the positive for the amorphous and the negative on the n-type silicon are both on the back. Without the usual contact fingers on the upper side, absorption is increased. In 2017, the Japanese Kaneka could announce an efficiency world record for silicon on its cell at **26.7%**.

Furthermore, Kaneka's modules at a bit over 24% were reaping the world efficiency record for modules so far held by SunPower.

For completeness, it should be mentioned that a number of other PV companies are proposing all-back-contact cells and modules as well, be it without the amorphous hetero-junction layer.

It is not the purpose to present here a full list of the various cell and module manufacturers with their products and conversion efficiencies that are on the market. There are many.

In a nutshell, typical commercial module sizes of today lie between 200 and 500 Watts. Efficiencies range from 15% to 20%. This holds also true for CIS, CIGS or CdTe modules that are offered on the world market.

There is a special market for single-junction GaAs solar cells. It is mostly for satellite applications and therefore very limited. Typical efficiencies of commercial cells here come at around 22%.

Even smaller is the market for multiple-junction cells. As they are expensive, only a small niche market emerged some time ago for PV concentrators, namely for optical concentration with lenses, but it was unable to develop very much. Multiple junctions employing various compounds of Ga, In, P, As, etc., have been developed by Sharp in Japan, Soitec in France, FhG ISE in Germany and NREL in the United States. Efficiencies between 44% and 46% at light concentrations of a few hundred Suns have been demonstrated on laboratory items.

6.3 R&D Attempts of Today

By now we can say the R&D job is done. At least for the key component, the PV module. Thirty $cents/Watt for a nice, reliable, product, what else do you want. I remember, long ago, in France someone from EDF said, PV will never make it even when the module cost becomes zero. Right now it costs not much more than the glass sheet on which it is deposited. What an achievement! Full turnkey PV systems are still a lot dearer, but this has more to do with soft costs and local policies.

What we did was the result of mass production development combined with R&D. In the glorious years at the beginning of the century when the markets exploded, global PV conferences exploded as well. Many thousands of people attended with enthusiasm.

Today, there is still an important global community left that is interested in R&D on PV. And it continues to be well connected. The world's biggest specialists' conference is the European PV SEC. I created the series in 1977. The 35th conference was held in 2018 in Brussels. The 34th conference was organised with the US and Japanese PV associations in Hawaii. I was one of its original initiators and stay associated with it.

These conferences are important for bringing people together, exchange latest results, and offer new ideas.

A highlight is always the part on silicon. Reviewing progress on structural developments, efficiencies and applications. The development of high-voltage modules of some 1,500 Volts and the corresponding inverters is also interesting.

Bi-facial silicon cells, once created by the Russians for space applications, were further developed by Spain's Antonio Luque, who created the company Isofoton in Malaga: They have also now made a new appearance.

Full-spectrum cells have emerged from time to time again, too. They go back to work in Germany since 1960, with contributions from **Klaus Thiessen** in Berlin and a discovery that I made myself on CdS crystals in my thesis in 1965, followed by some other researchers. A. Luque took up that subject again many years later, too, and took a patent on it.

The progress in alternative materials and structures is reviewed here. A highlight here is perovskite. Early expectations are far from being fulfilled so far. There is, in particular, a stability problem of the performances that go not much higher than 10% in efficiency on cells of a reasonable size.

Selenium is coming back a bit despite its modest efficiencies.

New multi-junction cells are being tried out. Efficiencies of 33% for III–V compounds on silicon have been announced—but for what market?

6.4 Looking Back: PV Discoveries, a World of Pioneers

6.4.1 The Discovery of the Photovoltaic Effect

In 1839, **Edmond Becquerel**, just 19 years old, discovers the photovoltaic effect: "...observes an electric current when one exposes unequally to Solar radiation 2 sheets of silver or gold in an acid, neutral or alkaline solution...". He conducted his experiments in his father's labs at the Muséum National d'Histoire Naturelle right in the centre of Paris. He was aware of the importance of his discovery and later wrote in 1867 the book *The Light, Its Causes and Its Effects* (in French). The building where this happened still exists. It is the same where his son Henry discovered in 1896 radioactivity. It was the radioactivity of uranium. Figure 6.2 shows the father and the son.

Later the Curies became very popular in Paris with their work on radioactivity that followed Henry's. There is an Institute Curie, a Museum Curie, Curie books, and Curie streets all over

France. Nothing named after Edmond Becquerel. His discovery was simply forgotten by the French. And I dare to say it was perhaps a bit more important for mankind's future than those of the Curies.

Figure 6.2 The Becquerels. On the right Edmond, the discoverer of the PV effect. On the left, his son Henri, the discoverer of radioactivity (picture credit: Loïc Babo, les Génies de la Science).

However, the English reminded it well. In 1989, on the 150th anniversary of the discovery, my friend, the late **Prof. Bob Hill** at Newcastle in England got the BBC to show a programme on Becquerel. Thereupon, I decided on behalf of the EU to create a Becquerel Prize for merits on PV. It is regularly remitted until this day in a formal sitting at each of the European Conferences PV SEC. Dozens of PV pioneers have been honoured since then. The first one was the late **Baron Roger van Overstraeten** from Leuven in Belgium—he was also a Belgian pioneer in informatics development, in Flanders as it were.

6.4.2 Towards a Practical Solar Cell

Selenium, and not silicon, was the first material used for a practical solar cell. Its spectral response to solar irradiation is less adapted than that of silicon, but it is easier to produce.

But before the first PV effect was discovered on selenium, the English **Wilboughly Smith** accidentally found a **photoelectric effect** on that material. The difference compared with PV is that the photoelectric effect does not involve the creation of a voltage, only the electric conductivity is altered by the incoming light as a function of its intensity. It is suitable as an optical sensor but not for the generation of energy. Accordingly, after Smith had published his findings in *Nature* in 1873, **Werner Siemens** in Germany produced the first photometer in 1874. Selenium optical sensors as light meters survived until the 1960s on German cameras.

A first PV effect on selenium was discovered by **Adams and Day** in London in 1876. It was also found accidentally as they had not built in purposely a diode that is essential for a PV effect.

The real thing happened in the United States in 1883 when **Charles Edgar Fritts** built a first selenium solar cell. It had a semitransparent gold contact layer that provided the diode effect. It is said that Fritts achieved an efficiency of just 1% with his cell. Interestingly, he is quoted as saying: "We may see the photoelectric plate competing with fossil fuel plants...". Fritts sent his cell to Werner Siemens, who showed it to the Royal Academia in Prussia. Everybody was impressed. Siemens was quoted saying: "first time the direct conversion of the energy of light into electrical energy". The first solar cells had too low an efficiency and were too expensive for practical applications. But still, the first electric car with selenium solar cells was built in the United States in 1912.

The world's first practical solar cell with 6% efficiency was built at Bell Labs in 1954 by **Daryl Chaplin**, **Calvin Fuller and Gerald Pearson** assisted by **Morton Prince**. It was a mono-crystalline silicon cell. In the meantime silicon metallurgy had developed sufficiently to make this possible. *The New York Times* wrote: "**Solar cells will eventually lead to a source of limitless energy of the Sun**". But for the time being first applications were for dollar bill changers. Later they became of strategic importance in the race between the United States and the Soviet Union with the Vanguard and Sputnik satellites in 1958. We will have an extra chapter on the technology rush of the silicon cells that followed.

Right from the early days when the understanding of semiconductors had sufficiently developed, one knew that many elementary semiconductors and compounds are suitable for PV. As a possible alternative to the just invented silicon cells came up then the "thin-film solar cells and modules". For their inherent properties, silicon cells must be at least some 100 microns thick. However, with other materials, achieving a factor 100 less is possible: the thin-film cells.

A good example is **GaAs**, which found a special market in satellite applications—because of its high cost, other markets are hardly accessible. A GaAs hetero-junction solar cell was first developed in 1970 by **Zhores Alferov** in the Soviet Union. At the same time, he developed the first semiconductor laser with this material—he is an exceptional personality who got both the Lenin Order and the Nobel Prize for his many achievements.

By now efficiencies up to some 28% have been demonstrated on GaAs solar cells.

In 1976, **David Carlson and Christopher Wronski** at RCA in the United States created the **amorphous silicon** (a-Si) solar cell. Previously **Prof Spear** in Scotland had found that the material is a semiconductor and can be n- and p-doped. You may remember the pocket calculators in the 1970s with their little PV strips. Those were of a-Si. Towards 2010, some 15% of the world's PV market was a-Si, but it did not last and has almost disappeared today. What broke its neck was an inherent degradation of amorphous cells. The main producer had been the Japanese Sanyo. Saving what could still be saved, Sanyo, which was acquired by Panasonic in 2009, developed the HIT cell. We described it in Section 6.2. In the HIT cell, a hetero-junction is produced by the contact of crystalline silicon with an ultra thin layer of a-Si.

Figure 6.3 St Petersburg, 2008. On the right, Zhores Alferov, Lenin Order and Nobel Prize winner, with Prof. Klaus Thiessen, Berlin, and the author.

Next to crystalline silicon, **CdTe** solar modules are nowadays the most successful ones on the global PV markets. The original pioneer for its development was the late **Dieter Bonnet** in Germany. For his achievement, he received a Becquerel award and a street was named after him in his hometown. Dieter established the basic technologies of the cell in 1971. It was the same CdTe/CdS hetero-junction that is still commercialised. Many people and companies got involved in Germany and France and in particular in the United States. There, an efficiency of 16% was achieved already in 2001 by **Ting Shu**, USF and Southern Methodist University in Dallas. Dieter started the first full production line for CdTe modules at the company ANTEC he had created in 1996 in Germany. It was the world's first fully automated PV module production line. The company did not survive the many ups and downs in the global PV industry. But **Harold McMaster** from Solar Cell Inc. in the United States came to see it. It certainly inspired him. He sold his company, which became **First Solar** in 1999. It is until now the global market leader for CdTe modules.

And there are **CIS and CIGS** solar modules. They stand for ternary and quaternary components of copper, indium, gallium and selenium. They had been proposed since 1971 as an alternative to silicon solar cells by **Prof Josef Loferski** from Brown University in Providence, USA. I remember well the meetings in the 1980s and 1990s where some managers of the US PV programmes expressed their anger that such complicated structures came up again on the forefront of research. But they kept attracting an enormous scientific and industrial interest and by now keep even a makeshift role next to CdTe on the global PV markets. And the winner is not one of those anticipated in the early days. It is Showa Shell Solar in Japan that is involved since 1993. With **Solar Frontier**, it runs large-scale production since 2007.

There is actually a lot more to be said about the fascinating times when PV started to put its neck out. Please refer to our book *Solar Power for the World: What You Wanted to Know about Photovoltaics*, published by Pan Stanford, Singapore, in 2013. Figure 6.4 shows the cover of the book's earlier version from 2010. In the book, 40 international pioneers of PV development report on their early work. **Morton Prince** who participated at the invention of the first silicon cell at Bell Labs in 1954 and later became the director of the US government's development programme of PV commented about the book: "I want to congratulate you on the quality of the book and the quantity of information that you were able to incorporate into it. Just reading the first 50 pages or so I found so much information that I was not aware of... And thanks for the tremendous effort in producing such a fine volume".

Figure 6.4 The first edition of the author's book *Power for the World* (2010).

6.4.3 The Silicon Solar Cell Development

The silicon world started in 1954 the year when the transistor and the solar cell had been invented simultaneously from that particular material.

Figure 6.5 Beach Party 2010 of *Power for the World* in Valencia, Spain, with some of the 40 book authors. My last picture with Hermann Scheer (second row, behind the two ladies).

In view of the enormous market perspectives for both, the electronic chips and the solar cells, the first thing that had to be developed was the large-scale production of **ultra-pure silicon**. The market for it grew indeed from its beginnings in 1954 to 600,000 tonnes in 2018, most of it for solar cells.

Silicon is derived simply from sand, which consists of silicon dioxide. One could employ the sand from the beach, but other cleaner quartzite deposits are used. After coke reduction in an arc furnace, the raw silicon must be cleaned substantially to become electronic-grade quality. Until this day, this is in particular done with the "**Siemens process**". The method was developed in the 1950s and patented by Siemens in 1973. The company Wacker in Germany adopted this process and created in 1968 Wacker Chemitronic, which later became Siltronic. Its director was the late **Werner Freiesleben**.

How to grow single crystals from silicon material was already well known since 1916 when Jan Czochralski invented the process for doing that. Hence, the door was open by then, the 1960s and 1970s, to enter the mass production of electronic devices on chips and solar cells on silicon single crystals.

Figure 6.6 A silicon single-crystal (picture by the author).

Next to single crystalline silicon, the multi-crystalline material for solar cells made its appearance since the early 1970s in Germany. The casting of poly-silicon blocks was developed by **Horst Fischer** and co-workers from AEG Telefunken and **Bernhard Authier** from Wacker. Both received in 1978 the German **Walter-Schottky Prize** for their work. Fischer, whom I knew well in those days, left then all the PV business and eventually ended up as vice-president of Siemens AG.

Multi-crystalline silicon looked interesting for lowering the production cost of solar cells. Hence, Freiesleben created in 1978 on behalf of Wacker Chemitronic the subsidiary **Wacker Heliotronic**, which is actually a programme! **In those days, Freiesleben became Germany's key promoter of the promises of a solar age.** He had very much influence on Hermann Scheer and encouraged him greatly when Hermann started his political initiatives on solar deployment.

However, Werner Freiesleben got neither support nor recognition from his company. After 30 years at Wacker, he left it in 1988 and Heliotronic became obsolete. Obviously, he had a conflict with his company that did not share his great vision on solar energy. And today Wacker Polysilicon makes big money in the solar business: At 80,000 tonnes, it is the world's second largest producer of silicon material. They should erect a statue in honour of their former far-sighted director!

Another challenge is the cutting of the silicon blocks, be they single-crystalline or multi-crystalline, into wafers. In particular, for solar cells, they must be the thinnest possible, between 150 and 200 microns. To this end, a new device was developed keeping the "kerf" loss, the powder loss during sawing to a minimum. It is the **wire saw**. A diamond cutting wire saw was developed by **Charles Hauser** in Switzerland with the support of inter alia **Guy Smekens** from the company ENE in Brussels and **Photowatt** in France. Hauser started as a consultant to **Solarex** and had achieved in 1986 his first multi-wire saw. He created the company HCT Shaping Systems and was able to sell it in 2007 to Applied Materials for $475 million. Later, the know-how was transferred to China. Currently Swiss company Meyer Burger is one of the leaders in wire sawing.

Finally, one had to address the optimisation of the solar cell technology, improving conversion efficiency and decreasing cost. The essential improvements in the standard p-n silicon solar cell were achieved by Comsat Labs in Washington DC and published in 1973. The work was directed by the late **Joseph Lindmayer**, a good friend of mine, who later created with **Peter Varadi** the company **Solarex**. The story of that essential development is reported by another friend of mine, the late **Martin Wolf** from

Princeton: "Silicon Solar Cell Development", page 113 in "Solar Electricity" proceedings of a conference held in Toulouse, France, in March 1976, of which I served as the secretary general.

First, Lindmayer's group developed the **Violet Cell.** It had 10% better efficiency than the conventional cell. This was achieved by three things: a better anti-reflection coating, a thinner diffused region on the surface and a finer grid-line structure. One step further, they approached the 18% efficiency with the **Black Cell**. Here, a process was introduced that has by now become standard on all solar cells that are on the global markets: The cell surface was shaped into a cone structure by an etching process. That reduces the optical reflectance by more than half. Additionally the photons penetrate the cell more obliquely thus increasing the long-wavelength absorption.

Figure 6.7 "Belgian" PV pioneers. They spent all their lives on PV. Brussels 2017. (Left) Guy Smekens, ENE in Brussels, (centre) Pierre Verlinden, now with Trina in China, and (right) the author.

A further improvement has been already mentioned: the **back contact solar cell**. Here both contacts, the positive and the negative one are put on the rear side of the cell doing away with the contact fingers on the front surface. It was proposed in 1994 in the United States by **Pierre Verlinden and Richard Swanson.** As it involves some higher complexity of the manufacturing process, it has not found general application.

6.5 The Early Vision of a PV Mass Production and Conquest of the World's Power Markets

6.5.1 US Pioneers Had a Dream

Imagine the power situation in the United States in the early 1970s. Billions of dollars had already been spent on nuclear power and hundreds of millions on coal power plants. It was planned to develop a liquid-metal nuclear breeder reactor and coal gasification for combined-cycle power plants by 1980. For the latter, one projected building one 5 GW breeder reactor every day. In the longer run, nuclear fusion was on the cards of the conventional power industry and their associated National laboratories. The energy Goliaths were flexing their muscles.

And solar power? It did not count at all. Twenty years after the country has invented the silicon solar cell, its yearly market for PV stayed at less than 0.1 MW in 1973, most of it for space applications. Yet today in 2018, the US market has reached some 10,000 MW. Something happened. In the early 1970s, the bottleneck for PV markets was the high cost of the solar cells: \$20/Watt for the tiny terrestrial market of those days and \$200/Watt for satellite applications. Today a silicon solar module costs \$0.30/Watt on the global spot market.

The wake-up call came from **William Cherry** of NASA: "The large-scale utilisation of solar energy will be a legacy for generations to come, something for all citizens to be proud of and a major step towards cleaning our planet both from a particulate and thermal standpoint. For these reasons, the large-scale utilisation of solar energy should be initiated. It just might be the difference between survival or the self-destruction of man".

Bill Cherry made this declaration in October 1971 at an assessment conference on the large-scale use of PV organised by **Karl Wolfgang Böer** at the University of Delaware.

At the same meeting, Josef Loferski, whom we mentioned previously, made the point that automobiles are a reference for mass production: In those days they cost $0.5 per kg of car. Solar cells came $6,000 per kg then and in terms of mass production similar considerations should apply. And that is what happened since then.

Figure 6.8 Karl Wolfgang Böer in 1973 in front of his Solar One in Delaware, USA. It was the world's first building with an integrated PV array (picture by the author).

A major event was the following workshop in October 1973 at Cherry Hill organised by the Jet Propulsion Lab. and sponsored by the National Science Foundation. My friend **John F. Jordan**, with whom I was later connected for his CdS solar cell factory in El Paso, made the point that the United States had at that time a global capacity of 400 GW of conventional power in place. And new conventional installations came at $250/kW, mostly financed by private investments. This cost figure came as kind of benchmark of cost against which PV had to compete in future.

The results of a panel on silicon cells were presented by **Paul Rappaport** from the RCA in Princeton, the later founder of the **NREL**. It proposed that by up-scaling, a cost of $0.10/Watt could eventually be reached at a production volume of 50 GW—and the probability of success were high: There is abundant material cheaply available, theory and technology are well understood, and reliability is proven.

It was an excellent projection. By now, some 45 years later, taking into account some 500% of inflation since the time the projection was made, we should stand broadly speaking at some $0.50/Watt today. We are even a bit lower on the global markets today, and the 50 GW are also exactly in the range of that market. Good work!

Only implementation did not proceed the way the early US pioneers had imagined it. They had demanded that the government start financing the necessary investments. "Indecision, fluctuating prices, political rhetoric will not generate the confidence for a privately funded PV programme", they said. But what always happens is that governments finance nuclear programmes but not the solar ones. As a consequence, the timeframe wanted by the US pioneers was not realised at all because the government did not follow suit. Until the turn of the century, the PV market in the United States did not really move.

What moved then first was the German market. Hermann Scheer had brought the social-democratic government in power there to open the doors to private investment with the FIT. The government in capitalistic America was supposed to finance the investments from the public budgets. That's a bit contrary to what you would have expected: that socialists invest from state budgets and capitalists encourage private business. However, America was catching up later and is in 2018 well ahead of Germany with respect to PV markets.

6.5.2 Europe in the Starting Blocks

There was a major event in July 1973: the International Congress on "The Sun in the Service of Mankind". It was organised by the French solar associations at the UNESCO House in Paris. It actually took place just a few weeks before the Arabs started the first oil price crisis—and before the American PV pioneers held the Cherry Hill workshop mentioned earlier.

The congress was attended by some 1,000 people from all over the world, the United States, Germany, the USSR, the Arab countries, etc. It was opened by **Pierre Auger**, the well-known semiconductor physicist (the "Auger effect"). He started his opening playing a hymn to the Sun.

I was asked to organise the PV part of that congress, and I assembled its proceedings.

Bill Cherry led the US delegation that contributed to the PV part of the Congress. The delegation from the USSR led by **academician N. S. Lidorenko** reported, in particular, on their work for optical concentration with PV.

The proceedings of this PV section of the congress were introduced with a welcome letter by **Wernher von Braun**, former director of the US Apollo programme. In this letter written shortly before his demise, he coined the expression **the coming "Solar Age"**. He remained farsighted throughout his life.

In 1974, consequent to the congress, I prepared a report under contract with UNESCO: "Solar Electricity, the Coming Energy Source". It summarised the prospects for PV for households, commercial purposes and central power stations. I insisted on the opportunity for PV pilot plants for various applications and next to the development of better and cheaper solar cells, the need for a comprehensive PV system technology. Later, beginning 1977 when I had responsibility for the EU R&D on all renewable energies and PV, I implemented the programmes along those lines. Looking back now, it was right to insist early on the system aspects: Truly high-quality and cheap solar modules have well been achieved, but PV system costs still leave room for further cost reduction.

In the Appendix at the end of the book, we shall come back to some details of the EU programmes in the course of the critical years until the year 2000.

I am glad to note that we held in 1993 and 2013 congresses at UNESCO in memory of the 1973 solar event to review what was achieved in PV worldwide. The results were very satisfactory.

Fairchild Industries Germantown, Maryland 20767 (301) 428 6000

CLOSING ADDRESS BY W.V. BRAUN
presented by W.L. PRICHARD,
President of F.S.E.C.

July 1973

Gentlemen:

I wish to extend greetings to all those present at the international conference on Solar Energy. The use of the sun's energy has been an area of interest to me since the days of my undergraduate work. The solar industry is in its infancy today, much as the space industry was so many years ago when I first started dreaming of rockets to the moon. I believe we are at the dawn of a new age, one which might be called the "Solar Age".

The United States is actively involved in this effort to solve one of mankind's most pressing problems. The United States is building up to what is expected to be a $2 billion research and development effort to bring us into this new age.

Three broad applications have been identified by the recent National Science Foundation/NASA Panel as the most promising from technical, economic and energy quantity standpoints. These are: (1) the heating and cooling of residential and commercial buildings, (2) the chemical and biological conversion of organic materials to liquid, solid and gaseous fuels, and (3) the generation of electricity.

I am confident that solar energy can be developed to meet significant portions of our future energy needs, and, thereby help to avert the coming energy crisis. Solar energy needs the firm commitment of the world's scientific, business and political communities in order to harness its tremendous potential in a timely manner. In acknowledgement of this, my company, Fairchild Industries, has set as one of its goals the application of solar energy to large-scale terrestrial use and the commercialization of products involving solar energy utilization.

We at Fairchild look forward to this new technological frontier and to working with the international solar energy community. I hope the conference is successful, and, although I regret not being present, I know Fairchild and I will be ably represented by Messrs. Pritchard and Farmer.

Yours sincerely,

Wernher von Braun

Figure 6.9 Wernher von Braun, the architect of the Apollo programme bringing people to the moon. His letter from 1973 to the author coining the term "Solar Age", shortly before his death.

Chapter 7

The Wonder World of Wind Power

7.1 The Development of Global Wind Power Markets until Today

Some details of the global market evolution have been already given in the preceding chapters. They can be summarised like this:

Global power capacity of modern wind turbines for electricity production evolved from 18 GW of global capacity in 2000 to some 580 GW in 2018. Only some 20 GW in total are offshore by now. The yearly installation rate increased from 6.5 GW of new installations in 2001 to 63 GW in 2015, a tenfold increase in just 15 years. Over 5% of the world's electricity is provided today from wind turbines. It was 0% in 2000. In the United States, the generated electricity from wind powers an equivalent of 24 million households.

Further, $112 billion was invested in 2016—the year for which consolidated figures are available—on 63 GW of new wind power capacity. It includes the investments in 2.2 GW of new offshore wind parks that are costlier to build than onshore parks. The figures show that nowadays turbines still come on average over $1,500/kW installed and turnkey, all costs included.

The Triumph of the Sun: The Energy of the New Century
Wolfgang Palz

ISBN 978-981-4800-06-8 (Hardcover), 978-0-429-48864-1 (eBook)
www.panstanford.com

In practical terms, for the supplied electricity, there are a lot of different prices applied in the various markets around the world, from some 9 $cents/kWh, the last EEG payments in Germany, to 4.9 $cents/kWh for the newest offshore park at Kriegers Flak in Denmark, and 3 $cents/kWh or so in Mexico, Peru, Morocco and Egypt. The United States announced that by 2021 no support to the markets might be needed, and Germany's latest auctions give this perspective even for the more expensive offshore parks by 2022.

The top countries for new wind power installation in 2016 were China, followed by Europe, the United States, Germany (when counted apart from Europe), India, Brazil and France.

There is a noticeable difference between the growth of wind power and PV over the last few years: First, the global wind power markets reached already their cruise routes at ±60 GW of new installations per year by 2013 while PV markets exploded since the same year on a new scale and came to a cruise route on their own only since 2016 at 70 to 100 GW of new installations per year.

Second, Europe has not been sidelined in the global wind power markets, as it was the case for PV. Europe's wind power market remains robust—although new challenges are foreseen in the immediate future towards 2022, in particular in Germany, Europe's market leader, to overcome new market constraints introduced by the auction system and the obligation to further improve technology and reduce cost.

Third, European wind power industry still leads the world, contrary to PV, where China has taken over: Companies from Denmark, Germany and Spain were able to keep all their strength previously developed when the first markets appeared at the end of the last century.

The future looks interesting with a new market environment and newcomers. In 2017, for the first time Russia has entered the global wind power scene with an auction on its first 1.9 GW.

Shell announced in 2017 that it thinks big on wind power: They see 200 GW of new offshore, nothing less—as mentioned before, today's global offshore market stands at some 20 GW, cumulated.

And there will be new technological opportunities.

7.2 What the World Achieved in Wind Power Technology

7.2.1 How It All Began

For centuries, windmills had been used for water pumping and grain grinding. Only towards the 1890s, the time when the first electric power plants were built by Edison, it was also time for the first wind power turbines to emerge. In 1887, James Blyth in Scotland built a windmill with sails that produced electricity for his cottage: the world's first house lighted with wind power. In the same year, Charles Brush in Ohio built a wind-driven generator with a dense range of wood blades. It is reported that it worked for 20 years. With the many blades, it was a low-speed turbine with a low efficiency.

The Dane **Poul la Cour** is considered the inventor of the modern type of wind power generator. He built his first turbine in 1891. A century later, the European Wind Energy Association (EWEA) created a Poul la Cour Prize for merits on the subject, and I personally was one of those who got that award at one time.

7.2.2 Today's Wind Turbines

The turbines of today look elegant and are efficient, but they are robust, too. When hurricane Harvey passed Texas, Florida and Georgia in 2017, not one of the many turbines that are operating there had been damaged. The industry is working to perfection: Projects are completed on time at no cost over-run. The industry masters a technology that is demanding: A large modern turbine is composed of up to 30,000 pieces. Transport of tower and blades can be an exploit.

The average turbine power is approximately 3 MW today. In the early 1980s, when the first big market started at the Altamont Pass in the United States with 16,000 machines installed, the individual turbines had a power of 100 kW only; one had not fully understood the technology in those days. Three megawatts is a kind of theoretical optimum for onshore applications. For offshore machines, much larger machines are suitable taking into account the additional criteria of operation and maintenance in the hostile sea environment.

Figure 7.1 The Danish book *Wind Power* (2009), by the Poul la Cour Foundation.

Turbine towers come as a rule more than 120 m high, and towers even up to 178 m high are currently in operation. In 1995, a typical tower was not more than 58 m high.

Most rotors in use are over 100 m in diameter. Up to 138 m diameter rotors are in operation today. The Adwen 8 MW machine for offshore has even a 180 m rotor. However, note that Adwen was discontinued as a company by 2017 end.

There has been a lot of development: In 2000, 15 machines were needed for 10 MW capacity, against 4 today. Today's wind farms generate 50% more electricity than those built 10 years ago.

Today the world has some 200,000 large megawatt-size wind turbines in operation, and there are even more in the kilowatt-size running. Approximately a million of the small wind turbines are in use, most of them in China, but also in the United States.

7.2.3 Technology Developments

There is a basic difference between the technologies on offer on the global markets today. Major producers such as Vestas employ the more traditional gearbox connection between the rotor drive and the electrical generator. Others such as Enercon, Siemens and Goldwind have opted for "direct drive", doing away with the whole gearbox. The generator in this case is a "multi-pole generator" that delivers the frequency connection with the electric grid into which the electricity is fed.

There is still another version that differentiates Enercon, which employs exclusively wound field coils to produce the magnetic field on both sides of all the many poles, from the other manufacturers, which employ permanent magnets on one side. Rare earth elements are employed in modern machines on such generators.

The direct-drive technology made its appearance in the 1980s. In June 1988, I organised the big European Community Wind Energy Conference in Herning, Denmark. It was there that the direct-drive concept was presented to the general public. Inventors were **Herbert Weh** and co-workers of the Technical University of Brunswick in Germany. Weh worked on it since 1980. His paper was entitled "Directly Driven Permanent Magnet Excited Synchronous Generator for Variable Speed Operation". Figure 7.2 shows the generator they presented in their paper.

Till this day, the three technology versions have maintained themselves in parallel on the global markets.

Figure 7.2 The first "direct drive" generator of a wind turbine. Designed by Herbert Weh, University Brunswick, Germany (picture from Weh's publication from 1988).

An important focus of R&D is currently put on the development of larger "monster" machines for the offshore market. Vestas plans to offer a 9.5 MW turbine by 2020. In a few years from now, one anticipates 13 to 15 MW machines. Plans have already entered a concrete phase and test fields are being established in Denmark.

Concerning better and cheaper blades, LM, one of the leaders of global blade technology that was just bought by GE, is now working on a 69 m long hybrid carbon rotor blade with integrated lightning protection—with a "feather-weight" of only 20 tonnes.

Nordex in Germany already offers rotors with longer blades (diameter 149 m) that employ carbon as stiffening elements, too. Those blades are part of a new generation of 4 to 4.5 MW machines the company proposes for low-wind regimes. And who offers more?

Figure 7.3 The elegant 140 m-high tower of a modern wind turbine with its assembly crane.

The American GE offers 158 m rotors on 4.8 MW machines, also for low-wind regimes. Those blades that are provided by LM, its partner, employ carbon technology as well.

At the end of 2017, the world's largest wind turbine was installed in Germany near Stuttgart (at Gaildorf). Its total height is 246.5 m, with a 178 m tower plus the turbine. It is a 3.4 MW GE turbine. Interestingly, it is also being combined with a pumped hydro storage power plant that is ready end of 2018.

Much effort continues to be put on the turbine towers that are 200 m high and must be erected in very remote places. Steel or concrete towers or hybrids are in use and are being further developed. They are transported in pieces that are put together onsite or the concrete is poured in circular pieces that are put one above the other. Note the enormous cranes that are in use for putting the machines together in an environment that is often very hostile.

A sign of the wind power's current vitality is also the revival in R&D of old concepts that the world had almost forgotten.

Against the market trend, where all rotors employ three-bladed rotors, Dutch company 2-B Energy has developed the prototype of a "two-blader" that runs successfully since 2015. It has a power of 6.1 MW.

Vestas, the market leader, has built a "multi-rotor" machine as an R&D prototype. It has four turbines, each 29 m in diameter on one single tower, almost 1 MW in total. TÜV Süd has certified the machine.

And the High School in Würzburg/Schweinfurt in Germany has built a turbine with nine blades instead of the usual three. With the many blades, it is a low-speed rotor. Advantage: low noise. The efficiency turned out to be acceptable.

An entirely different approach is proposed by "airborne wind systems". They have in common to use kites held by cables several hundred meters high above the ground. One approach is to move an electrical generator on the ground, mounted on rails, by the mechanical power transmitted by the cable from the kite. Another one generates the electric power in the kite up in the air on airplane-like structures; the electricity is transmitted by an electric cable to the ground. The research has been going on for several years on such systems, particularly in Germany.

For the time being, only proof-of-concept structures have been developed. Enerkite in Berlin may have a 100 kW prototype running in 2019.

7.2.4 Some Industry Pioneers

One of the key pioneers in modern turbine development in Germany was **Alois Wobben**. He created the company Enercon, which is a market leader in the country until this day. He originally came from the University in Brunswick, where he got his background on power electronics. The university developed in the 1980s important know-how that he could use for developing his machines. It is the same that is used also for other well-known applications, e.g. the elevator. You may have noticed that modern elevators do not give a jerk anymore when stopped.

Enercon got its first development support from the programme I directed at the EU in Brussels. We also arranged the particular shape of the nacelle of his machines. It is a design by Norman Foster & Associates in London. Enercon had its first direct-drive turbine developed in 1993. I complained to Wobben why he did not employ permanent magnets, but he insisted.

From modest beginnings, Wobben made Enercon a highly successful enterprise. He was a genius entrepreneur.

A particular problem emerged with the United States. It concerned the variable-speed concept of modern turbines. Turbines running with variable speeds have to cope with lower mechanical forces than those running at fixed speed. All turbines in the world before 1939 used the "variable speed" design. And Enercon's had it, too. However, in the United States, U.S. Windpower took a patent on it in 1991, famous US patent 5,083,039. Eventually, the patent passed to GE when it bought the companies that owned it. And GE was successful in barring Enercon's access to the US market because of that patent. In a $1 billion anti-trust suit, Mitsubishi has accused GE of monopolising the variable-speed market. This controversial patent expired in 2011.

Coming back to the direct drive concept, one could also refer to our book *Wind Power for the World*, published by Pan Stanford Publishing in 2013. In it, **Friedrich Klinger** from Innowind in Saarbrücken, Germany, describes the success of the direct drive in general. Klinger set up in 1990 a team on direct drive at the University in Saarbrücken. He was one of the first pioneers and went to present his concept also to Enercon. Klinger's was based on permanent magnets. Later, a small company, Vensys, emerged from his research work. It still exists. In 2003, the Chinese Goldwind took a license from Vensys on the direct-drive concept. Nowadays Goldwind employs the direct drive on a large scale. It is one of China's largest turbine manufacturers.

Klinger worked since 1990 with Siemens, Erlangen, which developed a prototype. They offered the concept to Danish turbine producers, but they refused the offer. And Siemens gave up as well.

However, Siemens came back. Today, it is one of the world's largest producers in the wind business, and it employs the direct-drive concept.

Klinger also insists on the pioneering role played by **Hermann Honnef** in the 1930s in Germany. He had started to develop a 20 MW machine on a tower 500 m high with several 160 m diameter rotors together on the same tower. His machine also employed the direct drive. He conceived it for offshore applications. With the wind technology know-how acquired since then, Honnef's dream looks certainly less stupid than it perhaps looked when it was proposed.

7.3 Opposition to Wind Power

7.3.1 Not in My Backyard

Not everybody likes to have a wind turbine rotating in their vicinity, their residence, their community, or their workplace. There is the visual impact and there can be some noise. The industry learnt early on to mitigate the problem by involving the interested parties. As all those turbines produce a relevant income, financial participation in the projects was an obvious opportunity. Nowadays the communities of the areas were turbines are set up are in many cases taking a share in the investments and the profits. So are doing the farmers who are renting their land for wind turbine operation.

7.3.2 The Disco Effect

The effect occurs when a turbine's rotor interferes with the incident Sun rays, namely when the Sun stands at low angle on the horizon in northern latitudes or in the winter months. The effect can be annoying as I experienced once myself. Hence, when the risk of a later disco effect can be anticipated during project planning, one should better look for the turbine's installation site North of the living and working quarters, opposite to the Sun's visibility.

"Leave the Sun in the South and put the wind turbines in the North of the buildings".

7.3.3 The Bird Killers?

There can be a real danger of killing a major number of birds. However, such cases are restricted to special situations, when enormous swarms of birds encounter a wall of densely packed turbines. Examples are Tarifa in Spain near Gibraltar, where dense swarms of migrating birds cross over to or from Africa twice a year, or the Altamont Pass in California, where thousands of machines are running since the 1980s. It is proven that at Altamont many birds have been killed. The latter turbines were eventually dismantled in 2017.

However, it is a myth that wind turbines are bird killers in general. Germany runs a central data centre on bird casualties since 1986. As the country has some 28,000 machines in operation, the results are relevant. The available data suggest that in total 681 dead birds have been found near all the 28,000 machines. This is close to nothing. Extrapolating on the 200,000 machines in operation globally, one can estimate a number of up to 10,000 birds killed.

Come the ornithologists and correct, there are up to 100,000 dead birds a year, just in Germany. They did not find them because the dead birds have been eaten in the meantime by other animals. Sorry, this is pure speculation, no proof.

But even if it were true that each turbine kills 1 or 2 birds every year that would make 500,000 in total globally. Again, this is an unproven speculation. In fact, the global coal plants kill more birds than even that speculative figure for birds.

Going back to basics, one counts some 300 billion birds worldwide today. Cats eat 2.4 billion of them annually, just counting those in the United States. One billion of them, it is estimated, are killed crashing into windows or façades. Another billion are killed on high-voltage power lines.

Sorry, wind turbines cannot compete with such figures.

Chapter 8

Bio-Energy in Harmony with Nature

8.1 Earth's Biosphere

8.1.1 Evolution

Darwin is known to have discovered evolution in all living matter. However, actually there was the Frenchman **Jean-Baptiste de Lamarck** (1744–1829) who had discovered it earlier when studying invertebrates. He wrote books about it. The legend has it that he wanted to present one of those to Napoleon, but the emperor refused it and insulted the scientist, who broke into tears.

Specialists point out the difference between Lamarck's and Darwin's views. While for Darwin evolution followed natural selection, Lamarck thought that it followed opportunities: A giraffe develops a long neck to reach higher branches.

De Lamarck was also the one who coined the term "biology". He worked and lived as a professor at the "Jardin des Plantes" in Paris. And there the Parisians erected a big monument for him (see Fig. 8.1). This botanical garden in the centre of Paris is by now already 400 years old and has today a unique particularity: It is planted in order of the appearance of the plants in botanical evolution—starting with moss and ferns, the oldest up to the youngest, tomatoes, carrots, and roses. This latest plantation

The Triumph of the Sun: The Energy of the New Century
Wolfgang Palz

ISBN 978-981-4800-06-8 (Hardcover), 978-0-429-48864-1 (eBook)
www.panstanford.com

follows the tree of evolution as published in 2009 by APG III, a group of international botanists.

Figure 8.1 Monument for Jean-Baptiste de Lamarck in Paris. He was the first to understand evolution. In 2017, the author with Jodie Roussell (centre), CEO of the Global Solar Council, and Michael Eckhart (right), from City Group in New York.

And at this garden, evolution is written in stone. Figure 8.2 shows the tree of evolution as it is presented there.

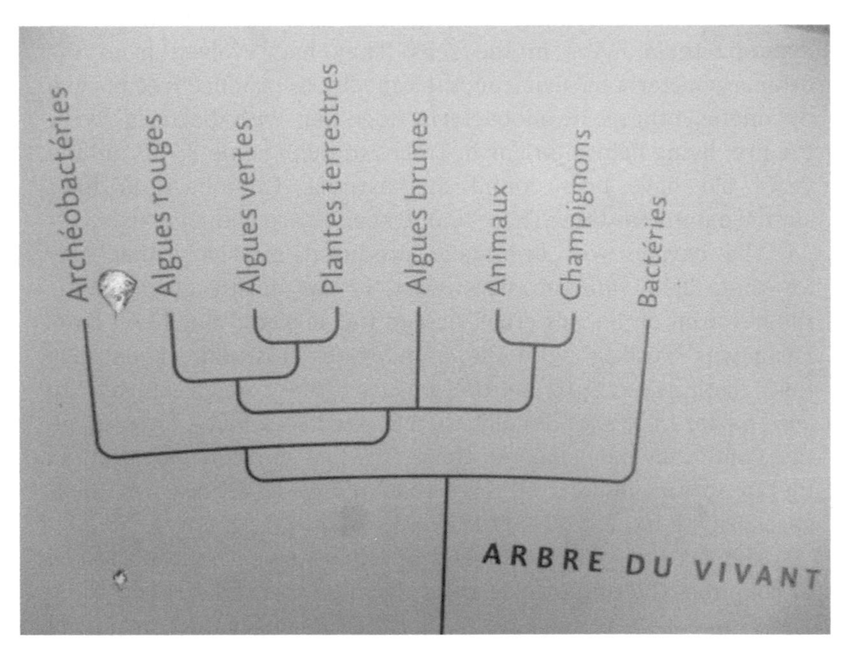

Figure 8.2 The tree of life's evolution, as shown on stone at the "Jardin des Plantes in Paris".

8.1.2 Origins of Life on Earth

All plants and animals as we know them had their common origin just 543 million years ago. By then, our Earth had already 8/9 of its life of 4,500 million years behind it. And a lot had happened to get at last the explosion of life—it is called the Cambrian explosion—started. Within just 10 million years, the ground was laid for the tree of life of all plants and animals, including us.

Today's organisms are "aerobic". They need oxygen to exist. However, at the beginning of times on Earth, the air was composed of nitrogen like today, but there was no free oxygen. That oxygen was all bound in water and CO_2. From there, it had to be freed first. That was achieved through photosynthesis by solar energy. In summary, it follows the following equation:

$$6CO_2 + 6H_2O + \text{Solar energy} = C_6H_{12}O_6 + 6O_2$$

The little unicellular animals that did the job were the **cyanobacteria** living in the seas. They had evolved from the ordinary bacteria by acquiring the capacity to produce free oxygen by photosynthesis. Cyanobacteria, together with bacteria, were the first living beings on Earth. Traces of them some 3,700 million years old have been found in Australia. Cyanobacteria have survived until our time. Their 10,000 species are known to us.

The oxygen was eventually produced massively that way for some 2,450 million years. First, it was mostly absorbed by the free iron on Earth's crust, first in the seas and finally on land. Earth was "rusting" with that iron oxide all around. Eventually that rusting came to saturation and free oxygen started to accumulate in the air. Finally, then started 543 million years ago the Cambrian explosion. At times, the oxygen concentration in the air in the millions of years that followed had reached 35%, specialists think. Today it is 21%.

The big bang of that biological explosion had one basic cell in common, the **eukaryotic cell**. That one evolved already some 1,600 million years ago from algae. Those algae had originally developed from bacteria and cyanobacteria. Better than the bacteria cells that do not have one, algae have a cell nucleus that contains their DNA. The eukaryotic cell evolved from such algae; and it has next to the nucleus one or several "**mitochondria**".

It is thought that those mitochondria were originally independent bacteria that were engulfed by the cells. The mitochondria have their own DNA that is independent from the DNA in the nucleus of the cell. The mitochondria are essential for the eukaryotic cell's life and death. They supply the energy to the cell. When the organism dies, the mitochondria start the decomposition process. That's how nature works today in all living organisms, except bacteria and viruses.

A human being as well is composed of eukaryotic cells, 1,000 billion of them, all specialising in certain functions. At the end of our life, the mitochondria start the decomposition process of the cells.

The most successful species on Earth were the dinosaurs. They survived for 170 million years. We, the humans, have still a long way to go to reach at least the first million years.

Successful reproduction is essential for survival. It is thought that sexual reproduction of eukaryotic organisms started already a billion years ago. It is seen as a powerful evolutionary force that does not exist in asexual populations, for instance, to increase chances of adaptation to the changing environment. Together with food search, sex is the fundamental driving force of life. Perhaps it contributed to the joy of life, too.

Plants have in their eukaryotic cells an additional element to capture sunlight: chloroplasts. Chloroplasts were inherited from cyanobacteria and have their own DNA, too. The energy gained through photosynthesis in chloroplasts from water with the release of oxygen is stored as the "ATP" material, which in turn is used to make the organic material from the CO_2 in the air—the whole is called the Calvin cycle.

8.1.3 The Biosphere of Today

Bacteria are the big survivors. They were the first living organisms on Earth and are still everywhere, some aerobe and some anaerobe. Their total mass may equal that of all plants, it is estimated. There are over 5 million species of bacteria. Actually, they are our friends (with exceptions). Without them, we could not live. Millions of them colonise our skin and our mouth. Billions live in our digestive tract.

On the other end of evolution, excluding such microorganisms, we got the more recent eukaryotes, the plants and animals. For both, the total number of existing species is estimated as 8.7 million (UNEP study from 2011): 6.5 million species are said to live on land and 2.2 million of them in the oceans. Most of them still have to be discovered, it is thought, and 1.25 million species are already in databases.

There are 7.7 million species of animals. The estimate for fungi is 0.6 million and that for plants stands at 0.3 million different species.

One counts over 60,000 species of different **trees**. According to the Food and Agriculture Organization (FAO) of the United Nations in Rome, today forests occupy 30% of the global land area. That was 4,000 million ha in 2015. Later figures of 2017 come to a higher estimate of 4,628 million ha.

There was a net loss of 0.08% of the world's forest area per year. The figure has decreased by half compared to the 1990s.

On an annual basis, there is now a loss of 7.6 million ha that is for half compensated by a natural gain through the growth of 4.3 million ha. The gains are observed in the forests of Russia, North America, and Europe. Deforestation on a large scale is going on in Latin America and Africa.

The biomass currently in stock in the world's forestry is 300,000 million tonnes. It decreased in the last 25 years by a total of 3.6%. The reserve is enormous even in comparison with the world's total recoverable coal reserve in the ground—leaving alone a consensus that it should stay there—that is only a factor of 3 higher. And biomass is renewable, while coal is not.

Ninety percent of today's plants are flowering plants. The 300,000 species of these plants that exist today stemmed from one single mother plant 130 million years ago. This follows from a study at the University of Vienna with scientists from 13 countries. That mother plant, it is found, was a hermaphrodite and the flower had three concentric petals like today's magnolia.

There are also 28,000 plant species with medicinal properties.

For birds, the American Museum of Natural History has recently increased its estimate to 18,000 species for a total population of 300 billion.

The number of Earth's mammal species is estimated to be 5,400 today. Just one of all these species dominates the world with a population of over 7.5 billion individuals: we, the humans.

In conclusion, what a richness there is in the world's biosphere! Only comparable to that of the Universe, just on a different scale: nanometres and microns instead of light years. The cosmos evolved from a big bang at one point in time and in space. So did the biosphere, by evolution from one bacterium with its primitive DNA some 3,700 million years ago.

8.2 Bio-Energy

Solar energy in the form of the Sun's radiation is converted by photosynthesis in Earth's plants to some 3,000 EJ (3×10^{21} Joules) every year. This big figure simply means that the solar energy captured annually as biomass tops the world's energy consumption of today by a factor of 10. However, it is not used that way.

The enormous amount of energy derived from the Sun as biomaterial is the basis of all animal feed and the food for us. But all that biomass, be it for productive use or not, is ultimately being decomposed again as it is not stored. In various ways, it is biologically or thermally recycled to water and CO_2, releasing the energy captured originally from the Sun, as infrared radiation, to space.

Bio-energy in the form of biomass in various versions is an energy giant, but just 1.4% of the total produced is so far employed as actual bio-energy, in the form of wood pellets, fuel liquids and biogas or simply as traditional fuel wood in the villages of the poor. Details have been given in some preceding chapters of this book.

There we saw that the modern forms of bio-energy developed speedily since the turn of the century. *Modern bio-energy is the leader among the world's renewable energies today.*

Unfortunately, on the other hand, the unsustainable use of fuel wood for cooking did not yet decrease as much as it should have.

It is true that modern bio-energy developed well this century, but there are huge margins for more.

Biogas is a point in case. The enormous flow of liquid effluents, in particular from husbandry, should better be recycled with the extraction of the energy contained in them. China, with its 42 million digesters in place, is an example, but even there no market saturation is in sight. Germany, the champion for biogas in Europe, is only at the beginning of its market potential. Its exploitation of full maize plants in digesters is an encouraging route.

Take the pellet market. It is in full swing, but it is much concentrated in only Europe so far. The global market currently stands at some 100 million m^3. This compares with the wood removal from forestry of some 3,000 million m^3 per annum, half for productive use in industry and half for fuel wood in the developing countries. As the latter is unsustainable use, it should be replaced anyway. In the nations of the Northern Hemisphere, with their considerable excessive growth of conventional forestry, the pellet market could well go far beyond the level of just 7% of all wood employed today for productive use.

And there is the market of biofuels for transport, ethanol and biodiesel. It made a flying start in the domestic gasoline markets in Brazil and the United States since the turn of the century. However, those markets reached saturation in 2008–2010. We mentioned earlier that there is a 10% "blend wall" now in the United States and an early market saturation in Brazil. However, there is no fundamental reason why bio-ethanol should not develop a lot more in future. I mean the classical one produced from cane in Brazil and corn in the United States. Different feedstocks are available at attractive prices and in greater volumes than ever before. And renewable bio-ethanol is attractive in terms of improving the environment and protecting the climate. The opposition is political. For years, environmentalists have propagated second-generation and third-generation ethanol, but they don't come. They are too cumbersome to produce, and there is no need for them. And today, guided by Elon Musk, one has discovered a new pet in transport policy: Dozens of billion dollars are now thrown by the industry under the applause of politics into the electric car business—even though people are reluctant to buy them even with big subsidies. And if Toyota, the world's leader in the car business, were right with their R&D on hydrogen cars?

Part 4

Power for the World

Chapter 9

Solar Energy for Survival Needs

9.1 The 1% Scandal

The explosive growth of the renewable energy markets in the new century occurred specifically in the industrialised countries, including China. As we saw in the preceding chapters, some 1,000 GW of wind power and PV have been installed there. Trillions of dollars were mobilised and 10 million jobs were newly created.

Virtually all systems were and are grid connected. The FIT, the tariff applied for feeding the clean electricity into the networks, was and is a key instrument for promotion.

The poor populations in the Third World, in particular those living in the rural areas were left aside. This became apparent in 2010 when the proliferation of PV reached new dimensions. In my public speeches since that year, I called it the 1% scandal: Less than 1% of all PV installed, or less than 1 GW went to the 1,000 million people in the rural world. It is an "absurdity" to bless all those who already get their electricity from the net with new clean power and miss the opportunity to connect the rural poor directly to the modern world—avoiding them to pass via a period of unsustainable supply of fossil and nuclear sources, like the industrial nations did.

It was a question of priority. There was nothing wrong actually in converting the existing energy system of the North to the

The Triumph of the Sun: The Energy of the New Century
Wolfgang Palz

ISBN 978-981-4800-06-8 (Hardcover), 978-0-429-48864-1 (eBook)
www.panstanford.com

renewables, which was due anyway, but the energy poor deserved the priority. However, it did not work that way.

9.2 The Problem

An overview of the general situation can be found in the World Bank's "2017 State of Electricity Access Report".

The first finding is that still 3 billion people rely on solid fuels for cooking and kerosene for lighting—unsustainably.

In 2014, the year for which figures are available, 1,060 million people had still no access to electricity. Yet in the last two decades, 1,700 million people were added to networks, in particular in urban areas. The number of people without access to electricity decreased everywhere except in "Black" Africa. There, it increased from 480 million in 2000 to 609 million in 2014. The reason was that population growth was higher than the rate of electrification.

Next to Africa, India and Bangladesh have the highest rates of non-electrification. Their governments do a lot to improve the situation. India is famous for the electrification programmes for its villages. However, it has been noticed that electrification often benefits public spaces only and leaves 90% of village families in the dark.

In September 2015, the UN adopted its 2030 Agenda for Sustainable Development. It recognises the key role of energy for sustainable development, but a World Bank analysis makes it clear that 100% electrification for all by 2030 is not on the cards.

9.2.1 Financing

Many countries of the Third World lack the necessary resources to develop their infrastructures and, in particular, a reliable supply of energy and electricity. The most forgotten ones are, as per the definition, the poor in the rural areas.

Development aid is a key for contributing to the lacking finance on behalf of the "rich" industrial nations and the related international institutions, such as the World Bank. Those aid budgets are considerable in size. They have even increased by 66% since 2000.

The EU Commission in Brussels is the world's largest donor—its aid is not a credit, it is non-reimbursable. I was an official there in the "EuropeAid" service in charge of it for some years. The EU Commission's aid programmes started in 1957. Currently they amount to some €7 billion per annum with priority to "Sub-Saharan" Africa. An official priority of the Commission's programme is the eradication of poverty and sustainable development.

In addition to the Commission's aid budget, the EU member countries provide an aid budget on their own. The dedicated budgets of the United Kingdom, Germany, France, Denmark, etc., are considerable, too: Altogether, Europe spends some €55 billion of aid per year. The United States comes second followed by Japan.

Financing is not part of the problem why sustainable development in the Third World is late.

9.2.2 Sticking to the Energy Options of Yesterday

In the industrialised countries, the electric utilities had to learn it the hard way that the time of the conventional power from fossil and nuclear sources is over. The traditional utilities in Europe, particularly in Germany and France, made in the last few years dozens of billions of Euros of deficit, and their stock market value fell dramatically; they had to reorganise themselves to become renewable energy utilities.

As expected, the energy business in the Third World took more time to follow suit. Until now, it was engulfed in the traditional way of proceeding: fossil fuel power plants and a network of grid lines all over the countries—at best, some mini- and micro grids employing diesel engines to serve an autonomous area of up to some 50 km across. The main beneficiaries are the cities and towns, and the villages are by and large left aside. That was the way the electrification rate in the developing world was dramatically increased the last 20 years or so.

Africa currently employs 42% of oil for its energy needs, 28% of natural gas, and 22% of coal. It relies for 6% on hydro and only for 1% on solar and wind power. It is far behind the Northern nations that decided to go massively for solar energy.

Africa has a special "Clean Energy Programme", called the Africa Renewable Energy Initiative (AREI). It has a $10 billion budget by 2020 that is provided by the EU, Canada, the United States and Japan. It may be seen as symptomatic for the current situation when in April 2017 the head of the programme, an African, quit. He accused, rightly or wrongly, the European programme managers of focusing on big infrastructure projects rather than community-led solutions. And some projects are said to aid fossil fuel generation, some will be owned by Europeans...

9.3 A Result of the Problem: Migration

The military conflicts in the Middle East are the reason for the migration of people to Europe. So is the lack of access to electricity in the villages of the Third World. Europe was building a fence to stop the flood of possible immigrants and the US president is keen to build a wall against immigrants from Latin America.

As the young people in villages see no perspective for themselves at home, they go to cities, a reason why over half of the world's population lives in cities by now. Life in cities is difficult for the newcomers, so these young people try to escape to the nearest developed countries—a problem for the migrants who may end up in misery and a political problem for the developed countries that are keen to protect themselves against unlimited immigration.

There is no easy solution to fight the problem. However, one option has not been tried hard enough: deploying large-scale electrification in rural areas. When it is done with solar energy, it is not only done sustainably but it provides locally the much needed jobs and business opportunities.

Refugees of the war in Syria are particularly hard hit. For a refugee camp in Jordan, Germany provided in September 2017 a 13 MW PV plant for light, food cooling and air ventilation.

9.4 Triumph of the Sun for the Rural Poor? A Glimmer of It

9.4.1 China, the Front Runner

China, the world's largest country, succeeded in providing full access to electricity for all in 2015. A great achievement!

The last families living in remote places, altogether 2.6 million people were provided with PV power between 2013 and 2015. On the government's initiative, systems of capacity up to 1 kW were centrally deployed, each providing full electricity: They are not just "solar home systems". The system costs amounted to some $3,200 each.

By 2018, China had some 800 MW of small wind generators for rural use installed, too. That means some 800,000 turbines at a capacity of approximately 1 kW each. They also supply several million people in the rural areas with clean electricity.

China's role in renewable energy deployment in its rural areas is above all visible in the field of bio-energy. In a previous chapter, it was reported that China has 42 million household digesters in its rural areas serving 160 million people.

Biogas production has a double function. In addition to the generation of energy, it develops sanitation development in villages. One must not think that hundreds of millions of people in the world's rural areas have access to "water closets". Ecological sanitation of faecal matter in digesters is a progress on its own.

Next to the giant China, neighbouring Nepal has been developing its biogas resources since the 1970s. I witnessed their efforts when I visited it in the 1990s. The potential in this small country is estimated at 1.5 million village digesters.

Generally, biogas has a major role to play in the rural world. Its further development, in particular in Africa, too, is a key to repress the use of firewood for cooking. The latter degrades the precious wood resources in the South, and the smoke released during cooking is very dangerous to health. The use of biogas is clean and fits with the development of ecological sanitation in villages.

9.4.2 The Spreading of "Solar Home Systems" for Survival Needs

The solar home system (SHS) is a PV device meant to provide a minimum of electricity to those who don't have any. For example, one can mention the one currently offered by French company Soltys; it is produced in Burkina Faso. It consists of a tiny 5 Watt PV panel, a little electronics, a 1.3 Ah battery, 3 LED lamps, and a

USB connection for a phone. It provides up to 38 hours of lighting. It costs just €20.

Modern SHS are indeed very cheap. In 1993, a PV lighting system cost $1,378 in Kenya, duties and taxes included. It had a 50 Watt PV module costing $340. Prices came down, in particular, thanks to the new availability of the low-consumption LEDs and their price drop. Their price was reduced by a factor of 10 in just 10 years.

Take the example of Bangladesh. In 2003, it had just 12,000 SHS in use. Their number increased continuously all the years and has reached 5 million.

In Africa, it is estimated that 60 million people benefit from SHS, a number exploding currently.

Figure 9.1 Kano, Northern Nigeria, 2011. The author with local promoters of PV and the representatives of the German Goethe Institute.

Above all, modern SHS are important for communication. In Africa, cellular phones are massively employed, also for health care and education. In "Black" Africa, there were 420 million cellular phones in operation by the end of 2016. Almost every second African had one. Thanks to the SHS, the rural poor benefit now as well from the service and not only those in the towns connected to electricity by the grid.

SHS in Africa help to bridge the digital divide.

9.4.3 A Trend for Larger Power Supply Systems

The trend for PV systems providing larger energy autonomy, thanks to the addition of batteries, is in full swing in Europe. Several billions of US dollars are being spent on the development of safe, reliable and cost-effective batteries.

The interest in electric cars that rely on batteries is not the only driving force for it. We saw in a preceding chapter that Germany has by now more batteries combined with PV in its recent markets than it has sitting in its newly sold electric cars. As a consequence of these trends towards mass markets, there is a sizable decrease in the cost of those electric batteries.

Li-ion batteries are preferred in the emerging electric car markets. However, the classical lead-acid type is still around, and to date, it is certainly the cheapest for applications in the power supply market in Europe and globally.

In Germany, for an average family, a PV array of 4 kW is estimated to cover a yearly consumption of 4000 kWh/y. This system costs some €6,000 when installed. For an 80% autonomy the client will have to add a 4 kWh battery or so. If it is of the Li-ion type, it will cost as much as the PV array, some €6,000. A lead-acid battery will be cheaper but may require more maintenance.

These recent developments in the economies of the North will no doubt benefit the electrification trends in the rural areas of the South after some time. Counting on a lower consumption of 1,000 kWh per family per year, the total investment cost may come in the range of €3,000 to 6,000. This is the up-front cost for a 20-year life system. Even bought at a usual credit rating for €250 per year, it may well exceed the finance of a poor family, but it provides an option for the central needs of a village, the schools, the hospital, entertainment, etc.

The best approach is to go for a PV system with battery storage in combination with a biogas unit. The biogas driving a combined heat and power unit, will definitely provide full power autonomy sustainably and at acceptable cost.

Chapter 10

Past Programmes Helping to Pave the Way

10.1 Solar Lighting by the Barefoot College in India

The Barefoot College, an NGO in Rajasthan under the leadership of **Bunker Roy**, played an exceptional role in the promotion of solar energy in India and Africa. It started its solar activities as early as in 1984. Its motivation was religion and ethics.

The college's particularity was to engage the local women for the deployment of the emerging PV plants in their villages: "grandmothers for solar energy". Women were trained as solar engineers. They manufactured thousands of solar lanterns and organised their installation. And it worked very well!

With the programme "Solar Electrification for Lighting", the college introduced PV for lighting in hundreds of villages, their schools and thousands of households. Some 500 kW of PV were deployed in total. Next to lighting, the college promoted solar water desalination, too.

The Triumph of the Sun: The Energy of the New Century
Wolfgang Palz

ISBN 978-981-4800-06-8 (Hardcover), 978-0-429-48864-1 (eBook)
www.panstanford.com

The college was active in the Himalaya region, in the Sahel countries of Africa and several other places in India and "Black" Africa.

10.2 Solar Water Pumping in the Sahel Countries of Africa

The Sahel countries south of the Sahara suffer naturally from the lack of water. Large areas are arid there. The EU Commission in Brussels started early on a governmental programme to improve the supply of drinking water and water for irrigation. It was a "regional" programme for the benefit of the nine countries of the Sahel region: Burkina Faso, the Cape Verde Islands, Gambia, Guinea Bissau, Mali, Mauritania, Niger, Senegal and Chad. The beneficiaries were—in the 1980s when the programme was started—the villages that had no access to clean water. In those days, 75% of the rural population was in that situation. Priority was given to villages with population between 500 and 3,000.

The regional programme was called PRS after its "French spelling". The idea emerged in 1985. Its first part went from 1990 to 1996. After extensive on-site evaluation of its implementation, the results were found encouraging enough to go for an extension that went from 2001 to 2007.

The programme was a solar programme and the technology employed was PV. In total, some 2 MW of PV was installed—the largest coherent volume of PV installation in Africa until the end of its completion in 2007. Almost all the 1,000 systems deployed went for the supply of drinking water. They were associated with devices for lighting and some food cooling. A very small number were used for irrigation.

The cost of the overall project exceeded some €100 million. It was essentially borne by the EU Commission in Brussels, which also had the general responsibility of the programme. I was involved in the programme from its beginning until it was extended.

Four million Africans benefited from this major PV water pumping programme. People not only benefitted from the devices producing the water but also were trained and educated on a more rational use of the water, maintaining a clean environment around the water points, health and sanitation.

10.3 Internet Connection to the Poor in Central America

At the turn of the century, when the Internet started its colossal triumph, its promoters realised that access to it in the Third World was a particular challenge. At that time, I was a member of an "inter-service group" at the EU Commission called **Bridging the Digital Divide**. I noticed that the IT experts in our group who did not know much about energy completely overlooked the problem of access to electricity: hence, no Internet without electric power.

An opportunity to do something about this arose when Fernando Cardesa, EU director in charge of the aid to Latin America, informed me that there was a new financing possibility for solar power in his budget. We agreed on a new initiative to provide access to both electricity and the Internet to some of the poorest villages in Latin America. Eventually the **Euro-Solar** programme was born.

The programme was officially adopted by the EU Commission in April 2006. It went on until 2012, with an overall budget of some €36 million, of which €7 million were contributed by the nations sharing the programme. It was a regional programme that was implemented via co-operation agreements of the EU Commission with the governments of Guatemala, El Salvador, Honduras, Nicaragua, Ecuador, Peru, Bolivia and Paraguay. The countries set up national implementation structures. All implementation steps were subject to international calls for tenders.

In total, 600 of the poorest villages—note that poor means poor in monetary terms; they are not poor at all culturally and socially—were the beneficiaries of the programme. Some 600 kW of PV for 600 systems was deployed.

The most innovative part of the programme was Internet connection via satellite. The power systems deployed consist of a 1 kW PV array combined with a small wind turbine and a gel lead acid battery. A small parabolic antenna provides the satellite connection, via the K band. The systems are installed in village schools and serve for education. Some systems are opened after school for the public, serving as "Internet café". They provide electric lighting, too.

The systems also opened the possibility to strengthen health care via the refrigeration of vaccines and medical products and facilitate operations through better lighting. Water cleaning via electric systems was also an option.

Euro-Solar was evaluated after completion by a group of independent experts that visited the plants on site. They remitted their report in 2014. (At this point, I thank my colleague Horst Pilger from the Commission for making the report and Figs. 10.1 to 10.3 available to me.) The report lists the many teething problems that were expected for such an innovative programme in a difficult rural environment. However, the overall conclusions are highly encouraging.

Figure 10.1 Joy about PV. Villagers in Peru at a fiesta about the acquisition of the Internet thanks to PV, the Euro-Solar programme (image credit: EU Commission).

Figure 10.2 School kids in Peru discovering the Internet: Euro-Solar (image credit EU Commission).

Figure 10.3 School kids in Peru (image credit: EU Commission).

The programme was a big success, in particular, for the villagers. The completion of the systems in their village was an invitation for them for an "all out" fiesta with much joy—perhaps the most wonderful popular success of solar energy the world has seen by now.

The programme is considered to have put hundreds of communities "on the map". Lack of access to power and the Internet is an indicator of social exclusion. The Internet is seen as a window on the world. Important is the connectivity with emigrated people.

As Euro-Solar was designed as a regional programme, it indeed provided the space for exchanging good experiences between countries. The results for health care turned out to be particularly good in Honduras. The connectivity was greatly improved in Nicaragua. In Peru, the programme has substantially contributed to spreading environmental concerns among institutions and people. In Guatemala, even wild replicas of the systems by churches were noticed.

The Organization of Ibero-American States for Education, Science and Culture drew inspiration from Euro-Solar when establishing its "**Light for Learning Programme**". It was decided in September 2011 by the ministers for education of its member countries. The programme had two themes, a foundation for energy without borders and "Ondula", standing for telecom and PV. Seventy thousand schools in Latin America, namely those for the indigenous populations in Argentina, Uruguay, the Dominican Republic, and others, adopted the systems inspired by Euro-Solar.

In Peru, Euro-Solar contributed to the design of the important programme "for massive rural electrification" and for alphabetisation. Peru has in place "Aulas de Innovacion Pedagogica" with a programme for Internet connection via satellite. In this case, the connection is kept free of charge.

10.4 The Satellite Industry to Connect the World's Unwired

There are only a thousand active satellites in space, against 2,600 that no longer work; 60% of these satellites are used for communication, GPS, telecom, weather, defence and agriculture.

Today's satellite business is enormous; globally, it amounted to $260 billion in 2017 and is expected to grow five times by 2030. Further, a business of $128 billion is associated with the services mentioned before, namely communication. Only a tiny number is currently employed for providing broadband Internet. Yet the number of those lacking Internet access today is estimated at 3 billion people. The terrestrial Internet networks reach only 10% of the population in the Third World.

One of the problems associated with satellite connection for Internet access is its high cost—while TV reception from satellites is free of charge. The cost was a major hurdle when our Euro-Solar was designed, as the connection cost is too high to be borne by villagers. However, the situation is changing. Spanish company Quantis, a small Internet provider in association with Hispasat, claims that it charges today €30 for 22 megabits/s, a tiny fraction of what it used to be at the turn of the century.

The business of connecting the unwired looked like an enormous business for the industry, a golden age. In 2015, big shots such as Elon Musk and Mark Zuckerberg invested in it. Crisis followed a year later when Intelsat and Eutelsat went down the drain with their share price. In the 1990s, others already had got their fingers burnt. In those days, a company called Iridium spent $6 billion on "one world, one phone". Six months later, they were bankrupt. A similar disaster happened with O3B in 2007 on "broadband to the other 3 billion".

However, Iridium came back. End of 2017, the company, which specialises in telephone connection via satellite, had 40 satellites in space. Iridium is associated with Musk's launcher company SpaceX to spread a network of 81 satellites in total. Cost: $2 billion.

Eventually, however, things tend to turn around again also for Internet connection.

The small service provider Quantis makes good business as it addresses the market of the 20% of Spaniards still unconnected, and those in Morocco, Chad, etc.

Eutelsat had two satellites placed in orbit in 2017, one above the Pacific with its capacity sold to Panasonic, and the other, the KA-SAT, above Europe. Both are in the geostationary orbit and provide connection to the passengers on airplanes.

The year 2017 was also the start of a "pharaonic project", the **OneWeb**. It is a US initiative half owned by Airbus Defence and Space. Among the investors are Softbank, Qualcomm, Intelsat and Virgin. The objective is to put in space **900 satellites** for providing broadband, high-speed Internet access to "half of the world", as the company announced. It will be a low-orbit network, like the GPS. It will operate in the KU band at 12 to 18 GHz, providing 17 to 23 Gigabits per second. The purpose is to provide Internet access available and affordable for "everyone", homes, cars, trains, planes...

The first target, it has been announced, is to connect by 2022 all the 2 million schools that lack the access to Internet today.

The first 10 satellites are being built in Toulouse, France. The first satellite is supposed to be launched in 2018.

So there is hope that the digital gap may be filled in the next decade. Perhaps there will be the same progress during that period to give everyone access to electricity, too—solar electricity.

Epilogue: Life in Harmony with the Sun and Nature

First they ignore you, then they laugh at you, then they fight you, then you win.

—Mahatma Gandhi

We, the many pioneers of solar energy, went exactly through this cycle, and we got the announced result: We won.

After a tremendous intellectual and financial effort, the world has returned to its fundamentals. Our new century, the 21st century, is back to harmony with the Sun. We have adopted again the benefits of solar energy and turned away from nuclear and coal. Bio-energy, solar collectors, wind power and hydropower have become the spearhead of energy policy and investments since the turn of the century.

Our energy has become more decentralised. It has become less vulnerable providing higher security of supply. The massive deployment of the renewables creates trillions of dollars of economic value and millions of new jobs. It is the unique arm against climate change and pollution of the air, the land, and the seas. It is a new chance to get the left-asides out of poverty. Solar energy provides comfort for all in better living and working conditions with zero-energy housing and plus-energy buildings on the horizon.

Political decision processes of various kinds are at the origin of this ecological breakthrough. The overarching result is that solar energy and its associated energies have become economically competitive. They are on the verge of becoming the cheapest of all energies. Further, as energy is fundamentally important for our life, all of us together benefit from the process.

There is no reason yet to weaken our alertness. Politicians, especially those in the Oval Office, could be tempted to declare tomorrow a new nuclear programme or a nuclear war, the same way as they recently declared going for reviving the use of dirty coal, or going to Mars.

However, the new cost advantage of solar energy would it make quite difficult even for a conservative politician to turn the wheel around again against the Sun. Earth in harmony with the Sun has the better arguments and a much longer life than a few trouble-making politicians.

Appendix

A.1 Getting Europe Ready for the Solar Revolution

Nothing can be created out of nothing.

The world has seen important R&D programmes on solar energy going on in preparation of its large-scale implementation—that started exactly with the turn of the new century, as described earlier in this book.

In Europe, a relevant development effort started in the 1970s under guidance of the EU Commission in Brussels, and I was the official in charge of it. The Commission provided contracts to joint developments in industry, universities and specialised institutions throughout Europe. The programmes were implemented in official advisory committees with the authorities of all EU member countries. There were 12 member countries in those days: Germany, France, Italy, the United Kingdom, Ireland, Spain, Portugal, Greece, the Netherlands, Belgium, Denmark and Luxemburg. Projects were funded from the Commission's own budgets with contributions of the national budgets of the countries concerned with a particular research.

The EU had in those days as well an official co-operation agreement with the solar programme of the US Department of Energy that I had arranged via the US Embassy in Brussels.

Contacts between specialists were encouraged by the EU Commission in numerous contractors' meetings and in dedicated international conferences on PV, wind power, bio-energy and solar architecture. The latter ones were attended by thousands of people from all over the world. Well-known examples are the European conferences on PV that I initiated in 1977 on behalf of the Commission. It is a series that still goes on. Conferences were early on an important tool for dialogue also with non-European activities. In the field of PV, those contacts were particularly important with Japan, which always had been a world leader in the field.

A.1.1 The EU Development of Solar Energy since the 1970s

Regular R&D programmes were set up and implemented by the EU Commission in 1977 under my direct responsibility. Early on particular strategies and guidelines were adopted for them, endorsed, together with a budget, by the European Council of Research Ministers. They were implemented by official calls for proposals. Financing was provided via contracts concluded and managed by the Commission in Brussels. The programmes had their heydays in the 1980s and 1990s.

A thousand contracts or so were concluded in these years and several hundred million Euros of finance provided for the development of solar energy. The results were made available in over 60 books published by the Commission in co-operation with commercial publishers.

The programme had five priority sectors: PV, wind power, bio-energy, solar applications to dwellings, and applications in agriculture.

PV had the highest priority of all budget allocation. Its highlight was the so-called "PV Pilot Programme". By 1983, a total of 1 MW of PV had been installed in dedicated plants. It was the largest PV capacity installed in Europe in these early days. The plants were ready-made for different applications. Most of them had battery storage, too. The largest plant of 300 kW capacity was installed on the island of Pellworm in Germany; 35 years, later it is still operational. Others served also for island and village power. One was for a school, one for a large TV emitter, one for an airport (Nice), another for a dairy farm, one for hydrogen production, and one for seawater desalination. The plants were installed in France, the United Kingdom, Italy, the Netherlands, Belgium, Greece, Ireland and, as mentioned before, Germany.

In those early days, much effort was put into device development, when the PV module cost was still 10 times higher than it is today.

Development of low-cost silicon wafers, silicon cell processing, encapsulation, alternative cells such as CdTe, CdS and CdSe were the priority subjects of research with industry and specialised institutes.

Figure A.1 The village Aghia Roumeli, island of Crete, Greece. Site of the PV Pilot installation, as seen on the left, from 1983 initiated by the EU Commission.

For wind power, a European assessment of its potential led to the conclusion that eventually the energy derived from wind could provide three times the total electricity need for the continent. That study was performed when its capacity had not even reached the first GW—one had to be farsighted.

The European Wind Atlas was prepared by Risoe in Denmark under contract with the Commission and so were the siting tools for turbine installations. The state of the art of technology in those days was established by studying hundreds of existing machines. Out of fundamental consideration, the optimal size of cost-efficient turbines was established as 2 to 3 MW. Until this day, it has proved to be an excellent guideline for the machines installed on land—the average size in Europe is exactly 3 MW today.

In 1985, we started the development of megawatt-size wind turbines. The programmes were called WEGA following the German acronym for "large wind machines". In WEGA I, three experimental turbines in the megawatt range were built in Britain, Spain, and Denmark; the Danish at Esbjerg was the largest with 2 MW. The Spanish one was the first to be erected at Cabo Villano at the northwest corner of Galicia—thousand others built there later by commercial investors followed. Then came WEGA II, which led to the development of the **first commercial MW-size wind turbines in Europe**. The Commission's contractors after tendering were Vestas, Enercon, Bonus (later bought by Siemens) and Nedwind. The total cost of WEGA II was €25 million, to which the Commission contributed €7 million.

For bio-energy, the EU programme involved the industry and agricultural research centres. The assessment of its potential included the interface with the "Common Agricultural Policy" (CAP) of the EU, the development of rural areas, job creation, production of new energy crops and recycling of residues and wastes.

The feedstocks under consideration were, besides forestry, agricultural crops such as *Arundo donax*, cordgrass, sweet sorghum and algae. Several million Euros were provided for the construction of four pilot plants in Germany, France, Italy and the United Kingdom for "methanol from wood" employing different gasification technologies.

In the field of solar applications for dwellings, the programme did not enter the proper development of solar heat collectors, but concentrated on testing and certification. It also got involved in the development of climatic data, thermal storage and solar cooling.

A highlight of the programme was "passive solar heating". *Passive Solar Handbook* for architects and engineers was put together and published. Passive solar components as well as simulation models and design tools were developed.

An assessment of the potential market of solar heating and cooling in Europe's building sector was established and published, too.

For solar energy applications in the agricultural sector, the programme started also with the assessment of its potential. It considered greenhouse heating, crop drying and increasing product quality, among others.

A.1.2 The EU Marketing Programme "APAS" of the 1990s

APAS is the French acronym for an EU programme in support of solar energy that stands for "Preparatory, Accompanying, and Supporting Actions" It was decided in 1994 when the European Parliament called for such a programme and provided extra €25 million on the EU budget line for renewable energies (RE). The programme was implemented via calls for proposals and managed by my division at the EU Commission. Over 340 proposals were received and eventually 70 contracts were signed. They involved over 300 European entities from industry and the electric utilities, architecture, research and academia, regional authorities, NGOs, etc. In fact, it brought together on a common platform some of the best competence available in Europe on the different solar energies in the 1990s.

The following overview is derived from the official EU publication "APAS Renewable Energies 1994, Project Synopses" under EUR 16876 EN 1996.

The following trans-national concerted actions involving different European entities were brought on the way.

A.1.2.1 On Route towards a Solar Energy World

- Promotion of RE in the European economy; the industry, job creation; bottlenecks and obstacles; cost-risk analysis
- Economic and environmental impact of a solar policy; the potential of a RE Europe
- Better land use, development of the local economy
- A geographical information system for large-scale solar integration
- Strategic planning
- Operational plans and policies for large-scale solar integration

- A decision support system
- An information network
- Education, training, master's courses
- Integration in the energy supply of communities
- Integration in the European supply infrastructure
- Network of European regions, accelerated regional integration, integrated planning in regions
- Energy packages for regions, municipalities, islands, creation of new structures
- Technology parks
- Legal, technical, administrative, structural conditions for electricity generation by auto-producers
- Networks for integration of RE for water production
- Electricity and water supply for socio-economic development in the Mediterranean countries
- Old coal mining sites

A.1.2.2 Urban Planning

- Energy conservation and sustainable cities
- **The Solar City**; the concept was developed in this APAS programme by the star architects Lord Norman Foster, Lord Richard Rogers, and Renzo Piano; subsequently, the solar town **Pichling near Linz in Austria** has actually been built; It has 25,000 inhabitants by now who have the privilege to benefit from a particularly friendly environment for energy and transport
- Towards zero-emission urban development; relationship between buildings, energy, the people, the micro-climate; the zero-emission town
- Technical and aesthetical integration of the RE in new settlements
- Urban planning; maximising the use of the RE
- RE strategies for European towns
- Stabilising GHG emissions in towns
- Collection of solar architectural data
- Use of daylight and natural ventilation in buildings

- Solar and bio-energy in small- and medium-size cities
- PV and thermal collector and envelope components for existing and new buildings
- **The Electric Home**, accelerated large-scale integration of PV in buildings

A.1.2.3 Regions outside Europe

- Southern cone countries in Latin America
- Decentralised rural electrification in India
- South African scholar solar systems
- Monitoring PV water pumping in West Africa
- Primary health-care clinics in remote rural areas
- Potential of desalination in the Jordan rift valley via its hydrostatic potential

A.1.2.4 Photovoltaics

- **Long-term large-scale market deployment (LSMD) of PV in Europe**; a study led by Bernard Chabot of the French ADEME in association with the British ETSU, the Dutch Ecofys, the University in Karlsruhe, Germany, and ENEL in Italy; **for the year 2030 the study projected in Europe a total installed capacity of 155 GW**; right now in 2018, we are not far from it, a nice work considering that it was performed when Europe had not even installed its first GW; The study assumed eventually a total market of 5 GW stand-alone systems installed, 50% of all residential roofs, each equipped with 4 kW of PV, 25% of all roofs of commercial buildings, and 10% of the total PV capacity installed on the ground; the study projected also much PV export that has not realised; on the contrary, Europe is importing most of the modules; and contrary to the projections, much of the market is actually devoted to systems installed on the ground
- PV for the world's villages, catalysing large-scale integration in the villages of the developing countries
- Implementation of solar home systems

- **Multi MW up-scaling** of silicon and thin-film solar cell and module manufacturing, MUSIC FM, a study by BP Solar in association with FhG ISE, ZSW, ASE, and Phototronics in Germany, IMEC in Belgium, Crystalox in the United Kingdom, and the Universities in Madrid, Lisbon, Utrecht, and Newcastle with many more subcontractors; the conclusion was that for a silicon production line of 500 MW per year, the module cost in the region of 1 Euro/Watt was achievable; no technological breakthrough was required; no unsolvable difficulties in manufacturing CdTe, amorphous-Si, or CIS thin-film modules have been identified either
- Removal of obstacles for PV technology

A.1.2.5 The Prospects for Bio-energy

- Large-scale biomass cultivation for energy in the EU; long-term impacts on farm income, employment, and the environment
- Technology, environment, land use, legislation, economic and social analysis
- Interface with CAP and GATT policies
- Incorporating externalities of biomass energy into the overall cost analysis
- Bioelectricity concept for large-scale implementation
- European bioelectricity network
- Municipal solid waste, garden residues
- Agricultural biomass for electricity and boiler fuel
- Bio-crude oil for engines
- Gasifiers
- Energy crops, *Cynara cardunculus*, *Arundo donax* (canne de Provence)
- Sweet sorghum for electricity and fuels in the sugar industry, *Robinia pseudoacacia* energy network

A.1.2.6 Wind Power

- Code of practice for the industry
- Wind-diesel desalination

A.2 Worldwide Travelling for Lecturing on RE

After leaving the EU Commission in 2001 as an official for reasons of age, I got involved in a worldwide promotion effort on behalf of the World Council for RE. The following presents a few examples.

Figure A.2 Farewell gift to the author in 2001 at the PV Conference in Munich when he retired as an official from the EU Commission. Provided by Peter Helm and signed on a Siemens panel by some friends.

A.2.1 Political Promotion of the RE

- Santo Domingo 2001, Sustainable Energy Seminar of the EU for the ACP Island States
- Beijing, 2005, World RE Forum, keynote
- Beijing, 2006, "Great Wall RE Energy Forum", co-organiser
- Beijing, 2008, International Energy Forum CIEF, keynote
- Beijing, 2009, China Energy Strategy Summit, organised by the Central Government, keynote at the Opening
- Washington DC, 2011, RETECH, keynote
- St Petersburg, Russia, 2008, Joffe Institute, Russian-German Conference, keynote
- Commonwealth Minister's Meeting, 2008; article in the reference book *Biofuels and the Global Food Crisis*
- Guatemala, Inter-American Development Bank, 2007, keynote
- Malaysia, 2008, RE Summit, co-organiser
- Lisbon, 2008, Global RE Summit IIR, co-organiser
- Dublin, 2009, RENEW A Roadmap for Ireland's Economic Revival
- Lecce, Italy, 2009, Festival dell'energia, keynote

A.2.2 RE on Oil and Natural Gas Conferences

- Algeria, 2007, Council of European Energy Regulators CEER, under patronage of the government, keynote
- Tehran, Iran, 2008, 13th International Oil & Gas Conference IIES, on the podium with Secretary General of OPEC El Badri
- Beijing, 2008, Conference of State Council with World Petroleum Council, keynote
- Baku, Azerbaijan, 2009, OSCE Energy Conference, keynote

A.2.3 Industry, Technology, Finance

- Beijing, 2009, at Great Hall of the People, 12th International Forum on Development of High-Tech Enterprises, VIP keynote

- Beijing, 2009, RE Entrepreneurs Club
- Vienna, 2009, European Institute of Innovation and Technology
- London, 2008 New Energy Finance Summit, invited as "thought leader"
- Oxford, 2006, Oxford Union Debate: "Britain goes for Nuclear", I opposed
- Tenerife, 2011, The Corporate Council on Africa
- Budva, Montenegro, 2009, Conference organised by GTZ. Meeting with Volkswagen manager W. Steiger to suggest a car engine market for CHP providing full energy autonomy in buildings

Figure A.3 Preparatory meeting of the International Renewable Energy Agency (IRENA) in Bonn in 2009. Organised by the German Government, directed by Hermann Scheer (picture by the author).

A.2.4 Buildings and Cities

- India, 2004, Conference on Intelligent Buildings

- Bahrain, 2005, Architectural Conference
- Beijing, 2006, Energy Policy and Mega-City Development: Towards 100% RE Cities, keynote
- Oxford, 2006, 2nd International Solar Cities Congress, keynote
- Barcelona, 2012, Smart Cities World Congress
- Munich, 2014, Future Cities Forum

A.2.5 Associations

- Global Bioenergy Partnership (GBEP) of FAO in Rome, permanent member since 2006
- Denver, 2002, American Association of Advanced Science
- ISES, contributing to many events
- IRENA, involved in its creation, namely at Charm El Sheikh, Egypt, meeting (2009) as a member of French delegation

A.2.6 Anniversaries

- UNESCO, Paris, 2013, 40th anniversary, "The Sun in the Service of Mankind" Congress, co-organiser
- Folkecenter, Denmark, 2013, 30th anniversary
- ITER, Tenerife, 2015, 25th anniversary, keynote

A.2.7 Adventures, Escaping Terrorism

- Syria, 2008: Invited by German authorities to the Damascus Conference of the Syrian Environment Protection Agency. Syria was then a normal, peaceful country; I felt like at home. I rented a car and drove 2,000 km across the country all by myself, all the way to Palmyra and the river Euphrates. Everything was quiet and pleasant. I had a major problem, not with terrorism but with a sand storm that almost cost me my life. It is unbelievable to see all the terror and misery, the hundred thousands of deaths, just a few years after that trip.

- Nigeria 2011: Also invited by the German Government, I went lecturing to Lagos, Abuja and Kano in the North. They got Boko Haram. Since then, Islamic terrorism has become more widespread and a lecturing trip is no more possible there today.

Figure A.4 Paris, UNESCO commemorating meeting of the 40th anniversary of the 1973 "The Sun in the Service of Mankind" Congress.

Figure A.5 Syria, 2008. The St Paul's Gate in Damascus from where the apostle escaped over the wall: St Paul was important as a founder of the Christian churches. Imagine what would have happened if he had not succeeded in his flight (picture by the author).

Figure A.6 Syria, 2008. The author at Palmyra. Most of it was destroyed when the IS moved in sometimes later.

Figure A.7 Syria, 2008. At the Euphrates, the cradle of civilisation. Shortly after this picture was shot, the IS made it their headquarters (picture by the author).

Balloon flights in China (picture by the author).

Index